STRUCTURAL STEEL IN ARCHITECTURE AND BUILDING TECHNOLOGY

STRUCTURAL STEEL IN ARCHITECTURE AND BUILDING TECHNOLOGY

IRVING ENGEL
Professor of Architecture
School of Architecture
Washington University
St. Louis, Missouri

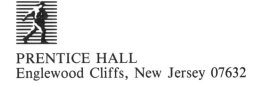

PRENTICE HALL
Englewood Cliffs, New Jersey 07632

LIBRARY OF CONGRESS
Library of Congress Cataloging-in-Publication Data

Engel, I. (Irving),
    Structural steel in architecture and building technology / Irving
Engel.
        p.   cm.
    Bibliography: p.
    Includes index.
    ISBN 0-13-854894-3
    1. Building, Iron and steel. 2. Steel, Structural.   I. Title.
TA684.E58 1988
691'.7--dc19   87-27781
                                                                    CIP

Editorial/production supervision and
   interior design: *Mary Carnis*
Cover design: *Diane Saxe*
Manufacturing buyer: *Peter Havens*

Cover photo courtesy of Elliott Littmann

Printed in the United States of America

10 9 8 7 6 5 4 3 2 1

ISBN 0-13-854894-3

PRENTICE-HALL INTERNATIONAL (UK) LIMITED, *London*
PRENTICE-HALL OF AUSTRALIA PTY. LIMITED, *Sydney*
PRENTICE-HALL CANADA INC., *Toronto*
PRENTICE-HALL HISPANOAMERICANA, S.A., *Mexico*
PRENTICE-HALL OF INDIA PRIVATE LIMITED, *New Delhi*
PRENTICE-HALL OF JAPAN, INC., *Tokyo*
SIMON & SCHUSTER ASIA PTE. LTD., *Singapore*
EDITORA PRENTICE-HALL DO BRASIL, LTDA., *Rio de Janeiro*

CONTENTS

PREFACE

When I first conceived the idea of writing a book on structural steel design, I wondered if there was truly a need for this sort of document. After all, library shelves at university engineering libraries are loaded with books on the subject. However, an evaluation of the available books on this subject (including classroom use of several) revealed that, with few exceptions, these books are written primarily for an audience of engineering students, at both the graduate and undergraduate levels. The few exceptions mentioned are generally of the "simplified" variety. That is, they are basically books that present procedures on a "how to" basis with little, if any, space given to derivations or principles that underlie the concepts. There are excellent books in both categories, considering their primary purpose and the audiences to which they are intended to respond. Indeed, I have recommended books such as these to practicing architects as references for their professional libraries, and I will continue to do so. However, when it comes to the university classroom, and dealing with students of architecture who are getting their first exposure to the subject of steel design, the available books, in my judgment, come up a bit short in terms of responding to the interests and concerns of this type of audience. In addition to students of architecture, I would place students in other building design professions, such as construction, construction management, architectural and structural technology, in the same category. In each of these fields of study, it is my opinion that students must be exposed to a reasonably sophisticated (and, indeed, academic) approach to the subject without being bogged down with material that is not relevant to their immediate interests in *building design and construction*. There is plenty of time in the life of people in these professions to look up a formula when this is needed. At the college level, however, scholarship should prevail and the basis for "formulas" and even empirical methods and techniques should be looked at very carefully. Where mathematical verification is available, this should be taught. Where empiricism guides decisions, the basis for this (historical or otherwise) should be analyzed, discussed, criticized, or supported, based on sound reasoning.

Therefore, the central purpose of this book is to provide students and practitioners in architecture and construction with the fundamental knowledge required for the design and development of a steel-framed structure. This necessarily includes several very important requirements. Among these are, at least, a rudimentary knowledge of the historical development of the steel frame (which is, essentially, a nineteenth-century American development), the procedural aspects of the design process, fabrication of the elements involved, an overview of the variety of structural steel systems, the effects of lateral forces on a steel frame, and the interfacing of various structural members. This book does not presume that people in these professions will actually perform the final analysis involved in the structural engineering of a building. Even for a small steel-framed building, this sort of analysis is best left to a qualified structural engineer.

Plan of the Book

The book includes responses to the requirements stated above, as well as the fundamentals involved in the computational aspects of the design of a steel structure. The computational procedures include a brief refresher of the structural principles involved. It is assumed that the reader has already been exposed to the analytical procedures involved at the principles level. In addition, design procedures for allowable stress design methods (or elastic design) for steel structures, as well as the newer plastic design methods for indeterminate structures (limited to continuous beams), are presented. Although the plastic design method is rarely used in the United States, the procedure is presented for study because it effectively points out the incredible reserve strength available in a steel structure and the factors of safety involved between the allowable stress level and failure.

The computational aspects are based on the idea that the student should understand the concepts that underlie the design process. The purely "how to" approach without the depth required for understanding the concepts is minimized, if not entirely eliminated. The noncomputational portions of the text, which are considerable, are included to provide the reader with a reasonable amount of knowledge of the building material being studied and a variety of concerns that the designer should be aware of, or address. It is hoped that the historical overview presented early in the text, however sketchy, will provide a sense of our structural heritage and will whet the appetites of students for more reading along these lines. Especially in the case of an architectural designer, some knowledge of the history of events that has led to our contemporary forms and philosophies is vitally important.

This book may be used for the basic learning process without supplementary references, such as the American Institute of Steel Construction's *Manual of Steel Construction*. The necessary data for wide-flange shapes is included in the Appendix.

In addition, reference is made throughout the text to various requirements of the specifications of the American Institute of Steel Construction (AISC). Although there are other specifications for the design of steel structures, we limit ourselves to the recommendations of the AISC Specifications.

The chapters dealing with computational procedures contain problem sets at the conclusion of the chapter. Other chapters contain recommended supplementary references. Many of these references are suitable for reading (as opposed to technical data or diagrams) and, indeed, it is recommended that they be read. In a classroom situation it is recommended that some research and writing be included as part of the learning activities.

Acknowledgments

I would like to express my gratitude to the Bethlehem Steel Corporation for allowing me to reproduce their technical data pertaining to wide-flange shapes, and to the American Institute of Steel Construction for permission to reproduce certain data from their publications. These permissions are noted in the appropriate locations within the text. In addition, I would like to thank Mr. Terry Zwick, General Manager of the St. Louis Division of the Bristol Steel Corporation, for providing many answers to questions related to the steel fabrication process.

Finally, my deepest gratitude to my wife, Margaret, for the many hours she spent as my chief editor for grammar, spelling, and punctuation, and as typist and word-processor operator. If it had not been for her assistance and dedication, this manuscript would still be on my desk in first-draft, handwritten form.

Irving Engel

Professor of Architecture
Washington University
St. Louis, Mo.

FOREWORD

Architects have always found delight or despair in the multitude of possible responses to an architectural problem. In our own pluralistic society such ambiguity also informs architectural curricula and at the same time enriches and frustrates the tasks of aspiring architects.

Understanding structures and structural principles introduces then a measure of very much needed clarity and rationality in the education of an architect. Professor Engel's textbook, *Structural Steel in Architecture and Building Technology*, is a delightful response to such a need. It also constitutes a precious contribution to the architect's primary task: that of form giving.

That this book is the product of many years of teaching at the School of Architecture of Washington University in St. Louis is an indication of its veracity. Also, it is certainly a source of collegiate pride for me.

Constantine E. Michaelides, FAIA

Dean
School of Architecture
Washington University
St. Louis, Missouri

STRUCTURAL STEEL IN ARCHITECTURE AND BUILDING TECHNOLOGY

Plate 1 Louisiana Superdome during construction. Courtesy of Irwin Buffet, New Orleans, La.

STEEL—THE MATERIAL

"The invention of steel and the elaboration of systems of construction based on its properties and those of its satellite materials, has in itself been responsible for the most spectacular changes in our social life."*

This statement was most appropriate when it was made, over 50 years ago, and the spectacular changes continue, being even more evident today. The continuing development of the quality of steel fabrication and construction techniques, and manufacturing processes leading to greater availability is, in large part, responsible for building systems that, indeed, have affected our lives. With structural steel (or steel-reinforced concrete) we have been able, only in the past few decades, to build increasingly taller buildings and wider clear spans. Many people work and live in "super" high-rise buildings that contain all of the facilities needed to serve even a great deal more than basic human needs. Can you imagine having your home, office, grocery store, restaurants, health spa, and so on, all within the enclosure of a high-rise structure? Save for walking the dog, one would rarely need to leave the building! Although such a life-style may not be acceptable, the point is that such "megastructures" would simply not be feasible without structural steel. The great "strength-to-weight" and "strength-to-cost" ratios provided by structural steel make such structures possible.

In addition to the great heights that buildings can reach through the use of steel, we can also enjoy the benefits that come from the ability to span long distances horizontally. Many of us can attend sporting events in a covered stadium during any season without being frozen, scorched, or rained upon. The long clear spans required

*Wells Coates, "Response to Tradition," *Architectural Review,* Vol. 72, November 1932, p. 168.

for such a facility would not be feasible without the light weight and great strength of steel (see Plate 1).

We know that structural steel is used for buildings of much smaller scale than those types suggested in the preceding discussion. There are a variety of reasons why steel might be chosen for the structural frame of a building, some of which have already been suggested. We will discuss the advantages of this material later in this chapter. For now, we concentrate on the material, its history, its properties, the variety of grades available, the cross-sectional shapes available, and an overview of the fabrication process.

AN HISTORICAL OVERVIEW

Although it is difficult to say when steel was invented (perhaps several thousand years ago), it is only since the latter part of the nineteenth century that it was developed for structural purposes. Prior to this development, cast iron and wrought iron were used in many building structures. Although these three materials are all related, they have very different properties. Let us, therefore, briefly discuss these three materials.

Cast iron.   This material is an alloy of iron, carbon, and silicon. It is cast as a molten liquid in a mold to develop the desired configuration. Cast iron is hard, brittle, and nonmalleable (not capable of being hammered or rolled into shape). Generally, cast iron contains about 2.0 to 4.5% carbon and 1 to 4% silicon. Because of its brittle character, cast iron has great resistance to compressive stresses, but low resistance to tension. Because of its chemical composition it has a high resistance to corrosion. Although cast iron was used for structural members (usually columns) during the latter part of the eighteenth century and through the latter part of the nineteenth century, it is used today for such things as manhole covers, machine parts, and other nonarchitectural applications.

Wrought iron.   Like cast iron, this material is also an alloy of iron in which many impurities are removed by the process of puddling. The process of puddling is one, essentially, where oxidizing agents are added to the molten mass in order to attract many of the impurities, which can then be easily removed. Wrought iron is malleable and shapes are made by hammering or, by some means, forcing into molds (hence the word "wrought"). Wrought iron has considerably less carbon than cast iron, and generally less silicon. Wrought-iron beams were used in building structures during the latter part of the nineteenth century since, unlike cast iron, wrought iron can resist some tensile stress. It was not uncommon, during the nineteenth century, to see building frames made of cast-iron columns and wrought-iron beams. Today, wrought iron is used primarily for decorative elements such as railings, gates, grilles, and so on.

Steel.   A tough alloy of iron containing carbon in amounts of up to 2%. It also contains other ingredients, whose amounts are very closely controlled. The careful control of the carbon content is what gives steel its unique properties. Steel is distinguished from cast iron by its malleability and lower carbon content. The property of malleability allows steel, under suitable conditions, to be rolled into a variety of shapes. In addition, but related, to malleability is the very unique property of ductility. Ductility means that steel has the ability to undergo large deformations without fracture. It is this property that makes steel a reliable and exceedingly dependable material for use in the structure of a building. One disadvantage of steel, as opposed to cast iron, is the fact that steel, because of its chemical composition, has a very

Plate 2   Train shed, Union Station, St. Louis, Mo., completed in 1894. An early example of a longspan steel structure. From *St. Louis Abandoned* by Elliott Littmann, copyright 1984.

low resistance to corrosion. Therefore, great care must be taken to protect exposed steel from a corrosive environment. In addition to ductility, steel is highly elastic, which means that it has a great ability to return to original proportions and shape after loads are removed. Steel is also unique among commonly used structural materials in that it is isotropic, meaning that it exhibits the same mechanical properties in all directions.

The events surrounding the development of iron and steel for building purposes and, consequently, the steel skeletal frame are varied and complex. To present a complete history of these events would, indeed, require an entire volume of work, at least. In this section, however, we discuss briefly some of the major events of the last few centuries that contributed to these developments.

Although iron and steel have been in existence for many centuries, these materials could not be mass produced. Iron could be smelted in small amounts and, consequently, its use was limited to small items such as hardware, and machine parts. The same is true of steel, which is an offspring of iron. Steel was a rare commodity whose use was primarily for weapons and other implements.

The Industrial Revolution provided the impetus for the development of mass-production techniques for manufacturing iron. Because of the large amounts of fuel required for smelting iron ore in large quantities, there did not exist a method for the mass production of iron until the mid-eighteenth century. It was then, in England, that the Darby family succeeded in using coke, instead of the scarce charcoal, for smelting iron ore. It was the work of Abraham Darby and the abundance of coal in England that led to the ability to mass-produce iron. This development was eventually to have an extraordinary impact on building construction techniques and the architecture of the years to follow.

Iron could be cast into a variety of shapes and, because of the newly developed ability to manufacture it in large quantities, new applications became evident. These included its potential for use in building structures. In about 1775 the first cast-iron structure was successfully completed by Abraham Darby the Third. This was a cast-iron arch bridge, across the Severn River, with a span of about 100 ft. The cast-iron arches were manufactured at the Darby's Coalbrookdale works.

From this point on a variety of structures in England and on the continent were built using cast iron. Initially, it was used mostly for roof structures where relatively long spans were needed, such as the Theatre-Française in Paris, which was built in 1786. Also, the use of cast iron became popular for use in buildings where timber roof framing would catch fire.

A significant step in the development of iron-framed building structures was the design and construction of a cotton mill building at Salford, Manchester, England, in 1801, designed by the engineers Boulton and Watt. The building is seven stories in height, which was remarkable for its time. It has an interior framework of cast-iron columns and I-shaped beams and exterior walls of load-bearing masonry. This building was, perhaps, the forerunner of the metal-framed "skyscraper" of true skeletal construction, which was not to come for many decades. The floor system, spanning between the iron girders, was made of brick arches and a concrete fill was placed on top of the arches to create a level floor. This entire system with its interior iron frame, exterior masonry walls, and brick arch floor supports was considered to be a fireproof structure. Fire had been a constant problem in mill buildings throughout industrialized England. In addition to the extraordinary feat, for the time, of erecting a seven-story building using slender iron supports, the use of I-section beams suggests extraordinary vision and intuition on the part of the designing engineers, since this shape is most efficient for members spanning horizontally.

In 1845, an engineer by the name of William Fairbairn employed a similar process of construction in an English refinery building, shown in Figure 1-1. However, Fairbairn introduced I-beams of wrought iron, in addition to cast-iron columns. This suggests the knowledge that wrought iron is less brittle than cast iron, and its use for members subjected to bending, such as beams, is more suitable. Fairbairn also employed the use of an arch form as support for the flooring material. Instead of brick arches, as used in the Salford Factory, Fairbairn used thin wrought-iron plates spanning between girders and bent into a shallow arch form. Concrete was used on top of these forms to provide a level floor. This system is, in a way, analogous to current practices in steel frame construction whereby metal deck is used and a concrete topping applied to form the floor surface.

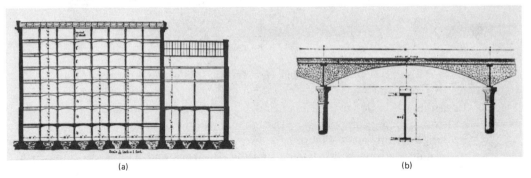

(a)   (b)

Figure 1-1 (a) English refinery, c. 1845, William Fairbairn, engineer; (b) English refinery ceiling construction.

While the work of Boulton and Watt and William Fairbairn in England laid the groundwork for the iron-framed structure, future developments and refinements in this sort of construction were largely American. These developments began in America in about 1850 with the work of James Bogardus and later, to some extent, Daniel Badger. The earlier work of William Fairbairn was an inspiration to James Bogardus, who was an inventor and innovative businessman, although he considered himself an engineer. Bogardus worked out a building system of individual cast-iron pieces consisting of columns and wall panels that could be bolted together. Using this system, he designed and built his own factory building in New York in 1848 (Figure 1-2). In this building, iron columns were substituted for masonry load-bearing exterior walls. This was, it seems, the beginnings of the true skeletal structure. While

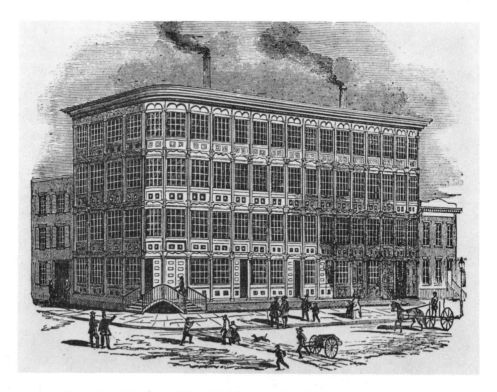

Figure 1-2   Factory building, New York, c. 1848, James Bogardus, designer.

Bogardus was a designer, Daniel Badger was a businessman and manufacturer of cast-iron building components, although many of the building facades were designed by Badger's iron works (Figure 1-3). From about 1850 through 1880 many buildings were built in a number of cities using the Bogardus and Badger developments.

While the stage was set, in America, for the development of the high-rise structure of true skeletal framing, there were structures of significance framed in iron being built in Europe. These include buildings of enormous scale, such as the Crystal Palace in England and the structures of the Paris Exhibitions of 1867 and 1889. These structures represent more the development of sophisticated engineering and construction techniques than a contribution to the eventual development of the material, steel.

In spite of the invention of the Bessemer process, in 1856, for the manufacture of large quantities of steel, the material did not come into general use in building structures until after 1880. Rolling mill practices had not developed and sizes and shapes of members suitable for structural use were very limited. In addition, the high cost of the steelmaking process made the use of the material economically prohibitive. It was in the field of bridge engineering, rather than building design, where it was recognized that the use of steel was necessary. Throughout the nineteenth century American railroad bridges suffered numerous failures. Many of these bridges were either cast or wrought iron, or a combination of both. It was finally realized that

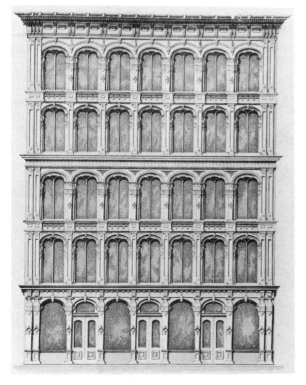

Figure 1-3 Design for a cast-iron store front. From *Badger's Illustrated Catalogue of Cast-Iron Architecture* by Daniel Badger. Courtesy of Dover Publications, Inc.

tne newer, more ductile, and more predictable material, steel, was necessary in order to avoid failures due to generally unpredictable and brittle cast and wrought iron. In architecture the use of steel was slow in coming, and, consequently, so was the full development of the high-rise skeletal frame. The mills that existed in Pennsylvania limited their output of steel to the manufacture of rails, but structural shapes were largely of iron. Through the demands of bridge engineers, the output of steel structural shapes began to increase. With this increase, and the tall buildings that were being built in New York and, especially, Chicago, attractive markets for the sale of this new material began to present themselves. ". . . . the Chicago activity in erecting high buildings finally attracted the attention of the local managers of Eastern rolling mills and their engineers were set at work. The mills, for some time past, had been rolling those structural shapes that had long been used in bridge work. Their own groundwork was thus prepared. It was a matter of vision and salesmanship based upon engineering imagination and technique. Thus the idea of a steel frame which should carry *all* the load was tentatively presented to Chicago architects."*

The Home Insurance Building in Chicago, built in 1883, was 11 stories in height and reputed to be the first "skyscraper" (see Figure 1-4). While the lower stories were framed with cast-iron columns and wrought-iron beams, steel beams were used above the sixth-floor level. This substitution for the original wrought-iron beams was

*Louis H. Sullivan, *The Autobiography of an Idea* (New York: Dover Publications, Inc., 1956), p. 312.

Figure 1-4 Home insurance
building, Chicago.

made during the construction of the building. The use of steel in an architectural
structure was a major event. However, while the mills were now rolling structural
shapes, the cost remained high. Consequently, until the beginning of the twentieth
century there were many buildings built of cast-iron columns and wrought-iron beams.
Eventually, more efficient techniques of manufacture and increased demand began
to bring economic feasibility to the use of steel for building structures.

With its great strength, predictability, ductility, and weldability, steel opened
new vistas for twentieth-century architecture. Essentially, the entire process started
at the Darby Coalbrookdale Works, in England. "Mass production of iron was now
possible, and this advance from manual production of the metal was to change the
face of the whole world."*

MECHANICAL PROPERTIES OF STEEL

Since the turn of the twentieth century, when steel started to become a common structural material, the methods of production, quality control, and the chemistry involved
have improved significantly. Consequently, the variety of steel that we use today has

*S. Gideon, *Space, Time and Architecture,* 5th ed. (Cambridge, Mass.: Harvard University Press,
1967), p. 169.

much greater strength. Also, there are, today, several grades of structural steel available to suit a variety of purposes. We shall discuss several of these grades shortly, but first it will be necessary to define the mechanical properties of steel, in general. For this purpose we will deal with the stress versus strain properties of a typical sample of mild steel. *Mild steel* is the term used for most of the steel employed in architectural structures. It is steel with the carbon content controlled and not greater than 0.30%. The most common grade of steel used today in structures has a carbon content of about 0.25%. There also exists hard steel, which is defined as steel with a carbon content greater than 0.30% but not greater than 0.60%.

Let's now refer to the stress versus strain diagram shown in Figure 1-5. It should be recalled from fundamentals that materials which obey Hooke's law have, when loaded, stresses directly proportional to strains, within the elastic limit of the material. Figure 1-5 is the generalized result of a test on a specimen of mild steel where strains (ϵ, in./in.) and corresponding values of stresses (*F,* psi or ksi) have been recorded.*

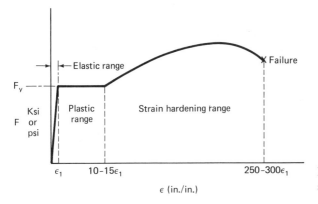

Figure 1-5 Generalized stress versus strain diagram—mild steel.

As shown in the figure, at some point there ceases to be a direct proportion between stress and strain. This point is known as the *proportional limit*. At this point, the specimen will begin to lose some of its elastic character. Consequently, for steel, the proportional limit is often referred to as the *elastic limit*. When stressed beyond this point the specimen will not completely return to its original shape and proportions when the load is removed.

The *yield point* is the level of stress beyond which there is a sharp deviation from the proportional relationship between stress and strain. In the most commonly used grade of structural steel this is the same, for all practical purposes, as the elastic limit. In all of the discussions to follow, throughout the book, we will use the term "yield point."

After the test specimen yields, there will be large increases in strains with, virtually, no increase in stress. The range of strains where this occurs is known as the *plastic range*. This range continues up to a point where the strain is about 10 to 15

*Ksi is the abbreviation for the term "kips per square inch," where a kip is the abbreviated term for "kilopounds," which is 1000 lb.

times the strain at initial yield. It is this property, which is unique to structural steel, that accounts for its ductility and reliability. If, in a structure, yielding occurs at a particular point due to overload, impact, shock, and so on, that part of the structure will accept no stress increase and stresses will distribute to other parts which have not yielded. Because of this property, sudden fracture will not occur as it may with a nonductile material, but the structure will show visible signs of distress.

Beyond the plastic range, there exists the *strain hardening range.* In this range the material again can take increasing stresses with nonlinear increasing strains. The test specimen will literally fail at strains of about 200 to 250 times the strain at initial yield.

The *modulus of elasticity* is defined as the slope of the line within the portion of the stress versus strain diagram where the material is stressed below the elastic limit. In physical terms the modulus of elasticity may be thought of as an indication of the strength of the material. Although there are a variety of grades available, structural steel is unique in that all grades of steel have a common modulus of elasticity of 29,000 ksi. Figure 1-6 shows the general relationship between several grades of steel, indicating that the slope of the straight-line portion of the stress versus strain diagram is the same, but yield points will vary depending on the grade of steel.

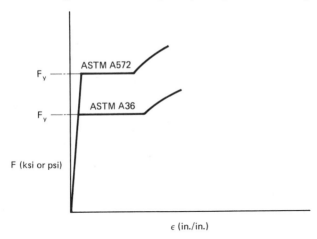

Figure 1-6 Initial portion of stress versus strain diagram showing constant modulus of elasticity (*E*).

GRADES OF STEEL

As suggested in the preceding discussion, there are a variety of grades of structural steel available. The various grades of steel are specified by the ASTM (American Society for Testing and Materials) designations. Some of these are described in the following discussion.

The most common grade of steel used in a building structure is ASTM A36. This grade of steel has a minimum guaranteed yield point of 36 ksi. While other grades of steel have higher yield points, A36 steel is the most commonly used because of the "strength/cost per pound" ratio. We will elaborate on this after describing several other grades of steel.

A higher-strength steel is designated as ASTM A572. This grade of steel has yield points ranging from 42 to 65 ksi, depending on the thickness of the material. The higher yield point is achieved by slight alterations to the chemical composition of the alloy.

In addition to the above, there is ASTM A588 steel. This is also a high-strength steel with yield points varying from 42 to 50 ksi depending on the thickness of the material. A588 steel is also commonly known as "weathering" steel, because of its unusual corrosion resistance properties. The increased corrosion resistance is provided by slight alterations in the chemical composition, most notably a slight reduction in the carbon content and the inclusion of a very small percentage of copper. Weathering steel may be desirable, architecturally, because it develops its own tightly protective oxide coating when exposed to the atmosphere. This coating then protects the inner parts of the steel from further deterioration. It may be said that weathering steels naturally "paint" themselves with a coat of rust, thereby giving a very earthy and, depending on aesthetic preference, a very desirable color to exposed steel structural members. Because of this property some savings may be achieved through the elimination of painting. Where painting would be desirable for preferred color, weathering steel provides a paint life much greater than can be obtained with A36 or A572 steel. One of the potential problems of which the reader should be aware is that during the initial development of the oxide coating on weathering steel, the oxides are not tightly adhered. Therefore, when rained upon, rust will run and may deposit stains on other architectural finishes. Care should be taken in areas where this situation may occur. This can be done through sensible detailing procedures. Some potential problem areas and possible preventive detailing measures are suggested in Figure 1–7.

The steel with the highest minimum yield point is designated as ASTM A514. It has a minimum guaranteed yield point of 100 ksi or 90 ksi, depending on the thickness of material. Normally, this grade of steel would only be used for columns in reasonably tall structures where the compressive stress is great. In general, there would be no advantage to using this grade of steel for a beam, because of the very high price/strength ratio.

There are several other grades of steel used, but the ones indicated would be most common in an architectural structure.* Let's now discuss several issues that would be considered in the determination of the grade of steel to be used for a given project.

It was mentioned earlier that A36 steel, while having the lowest yield point of the most commonly used grades of steel, is the most popular because of its relatively high strength/price per pound ratio. It should be noted, however, that there are oc-

---

*For a detailed list of available steels and their chemical compositions, see Bethlehem Steel, *Structural Steel Data for Architectural and Engineering Students,* Booklet 3025, current edition.

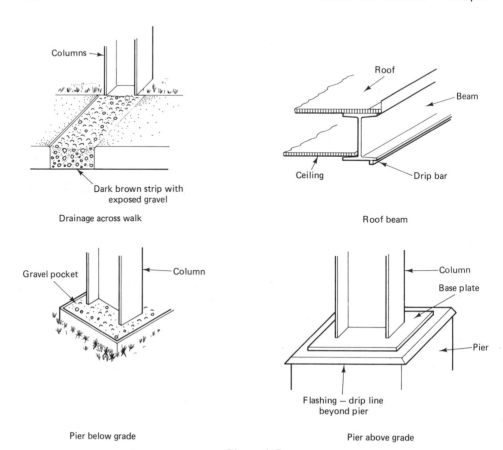

Drainage across walk

Roof beam

Pier below grade

Pier above grade

Figure 1–7

casions when high-strength steels may be used advantageously. One case, as mentioned earlier, is when a compression member must support large loads, such as a column in a tall building. The increased allowable stress that would be available with the high-strength steel means that a smaller cross-sectional area would be needed compared to that for a column made of the lower-strength steel. In fact, it may be necessary to use high-strength steel to avoid the necessity for fabricating a built-up column (one made of several shapes welded or bolted together). Where loads are very high, however, a built-up column may be the only way to provide the necessary cross-sectional area. When a structural shape is to be used as a beam, there is also the possibility for using high-strength steel, but such needs are not frequently encountered. For example, high-strength steel may be advantageous where the dead load of a steel frame must be kept to a minimum. The high-strength grade would lead to smaller beam sizes and, therefore, reduce the weight of the structure. An important point, when deciding on a grade of steel to be used as a beam, is that all grades of structural steel have the same modulus of elasticity. Therefore, a beam made of A36 steel and a beam made of one of the higher grades would both deflect the same amount under a given load. If deflection is an important issue, and it normally should be, the smaller beam sizes required through the use of high-strength steels may deflect excessively. It is for this reason, and an attractive strength/price ratio, that A36 steel is the more popular choice. However, where increased stiffness can be achieved through composite construction (see Chapter 4), there may be some economic advantages in using a higher-strength steel.

STANDARD SHAPES IN STRUCTURAL STEEL

Steel can be rolled into a variety of shapes and sizes. The most commonly used shape in a building structure is the wide-flange shape. Wide flanges are efficient members because they have large moments of inertia relative to their cross-sectional areas. The wide-flange shape should not be confused with the I-beam. In general, I-beams have a much smaller "width of flange/depth of section ratio" than that of a wide flange. Although wide flanges and I-beams are similar in cross section, they differ slightly in that the inner surface of the flange of a wide-flange shape is virtually parallel to the outer surface, and the inner surface of the flange of an I-beam has a pronounced slope. These cross sections are shown in Figure 1–8. There are other standard rolled steel shapes available. They include channels, tees, angles, and zees (see Figure 1–9). Angles may be of equal leg dimensions or unequal legs. Structural tees are cut from wide-flange sections.

These sections may be used for a variety of purposes. For example, the members in a steel truss are most commonly made from angle shapes. Angles are also used

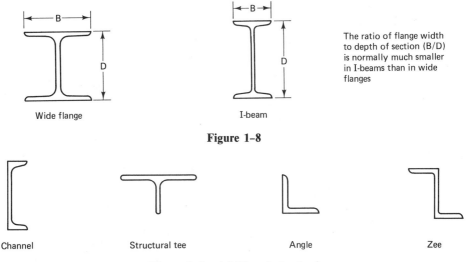

Wide flange   I-beam

The ratio of flange width to depth of section (B/D) is normally much smaller in I-beams than in wide flanges

Figure 1–8

Channel   Structural tee   Angle   Zee

Figure 1–9 Additional standard shapes.

frequently as lintels in wall openings of small to moderate width. Channel sections are mostly used for stair stringers, structural fascias, and purlins. Tees are sometimes used for top and bottom chord members of a truss or for structural lintels. Often, for a variety of purposes, these structural shapes are combined and a variety of possible cross sections may be developed as suggested in Figure 1–10. In addition to rolled sections, square, rectangular, and round tubing is available. Most commonly these shapes are used for columns, although, on occasion, rectangular steel tubing is used for beams.

The properties of all available steel sections are given in the *Manual of Steel Construction* published by the American Institute of Steel Construction. In the Appendix to this book, properties of wide-flange sections are given because our primary concern will be with these sections. The manner in which the various sections are designated will be discussed in Chapter 4.

| Wide flange with angle | Channel with angle | Wide flange with channel | Two angles | Two wide flanges |

Figure 1–10   Some possible combinations of shapes.

THE BASIS FOR CHOOSING STRUCTURAL STEEL

The purpose of this discussion is not to suggest that all building structures should be supported by structural steel frames. In fact, such a suggestion would be pure nonsense. Each structure must be evaluated independently by the architect and engineer in order to determine the most suitable structural system, given the variety of circumstances and conditions that may be encountered. However, it is important to note that there are certain advantages in the use of structural steel as the primary supporting system for a building. On the other hand, there are also some disadvantages to be considered. Let us now discuss both the advantages and the disadvantages of steel.

Some of the advantages of structural steel are:

1. High strength per unit of weight. This means that the dead loads of the building structure will be minimized. This may be important where foundation conditions are poor and total loads must be kept to a minimum. This is also an important consideration where long clear spans may be necessary.

2. The behavior of structural steel members is highly predictable because of the uniformity of the material, unlike other commonly used structural materials.

3. Steel has that marvelous property of ductility, which is, perhaps, one of its properties that makes it a very good choice, especially where a building may be subject to seismic forces.

4. Structural steel offers certain advantages when considering additions to existing structures. For vertical additions the construction procedures are relatively simple, provided that the original structure and its foundation have been designed to accept increased loads. When adding horizontally to an existing structure, such additions are also reasonably simple because of the ease of bolting or welding to elements of the existing structure.

5. The fact that most structural steel members are shop fabricated leads to a great amount of accuracy in the placing of bolt holes as well as speed of erection in the field. Actually, a steel structure may be thought of as a giant "erector set." All items can be cut to length, drilled, and have things welded to them in a fabricating plant. Because of this, a steel structure may even be designed to be disassembled and reassembled at another site. Because of the prefabrication of the parts, the construction phase of a steel framed building is much cleaner than for other types of buildings, where either formwork must be used or a great deal of on-site work is necessary.

On the other hand, steel has some disadvantages. Some of the disadvantages are:

1. Except for "weathering" steel, steel has a low resistance to corrosion. Consequently, when steel is exposed to a corrosive environment, high maintenance costs may be involved. In buildings where all steel must be fireproofed, this is normally not a concern since the fireproofing material will protect the steel.

2. Where a high fire rating is necessary, there would be an increased cost involved in the necessary measures to provide fireproofing for the steel. Steel is a poor material in a fire because it loses its integrity at relatively low temperatures.

3. While the tendency for a steel-framed building to have a relatively low dead load may not be thought of as a disadvantage, the light weight of steel can, in certain cases, be a problem because of vibration and "bounce."

In addition to the few advantages and disadvantages indicated, there are several other considerations involved in the choice of a structural system. These include cost (as always), availability, and transportability. The design of a structural system must be considerate of these realistic aspects.

FABRICATION AND ERECTION OF STEEL

Structural steel shapes are manufactured at rolling mills and shipped to steel fabricating plants where the pieces are prepared for a particular building project. The fabrication process is fairly complex and requires a number of exacting procedures. The fabricator of structural steel must first, based on the structural drawings, order the appropriate shapes in the necessary lengths from the steel mill. The fabricator must then prepare detailed drawings showing exact lengths to be cut, holes to be drilled or punched, items to be welded, and so on, so that the fabrication shop can prepare the pieces properly before they are shipped to the job site. Every detail must be shown and dimensioned accurately if the pieces are to fit together properly at the job site. The drawings that contain this information are prepared by the fabricator and are called *shop drawings*. These must be prepared in strict accordance with the architect's and/or structural engineer's drawings. A sample of a shop drawing is shown on Plate 3.

The erection process may be reasonably simple or complex, depending on the size and configuration of the building and the variety of equipment and techniques that must be employed. Essentially, the erection process is one whereby the steel members that have been fabricated are placed in their proper positions in the building frame. The erection of the frame may be done by the steel fabricator or may be subcontracted.

In addition to the proper positioning and securing of the fabricated pieces, there are several issues that must be considered. One of these is the manner in which the fabricated pieces will be shipped to the building site. The necessary routing of the pieces from the fabricating shop may suggest the maximum size of the various pieces. In some cases structural assemblies (such as large trusses) may be too large for shipping or handling. These would be fabricated in several segments and the final assembly would take place at the building site.

Site conditions must also be considered. This would include the accessibility to the building site of the necessary erection equipment. Also to be considered would be the possible effect of the proximity of surrounding buildings on the manner in which the steel frame is erected. It may also be necessary to develop a schedule for the erection process, coordinating with other trades involved in the construction of the building. Consequently, an erection scheme may need to be developed simultaneously with the fabrication process.

Once the fabricated pieces are delivered to the site and the erection process begins, there are other concerns that are the responsibility of the steel erector. Primarily, this would include the placement of any temporary bracing that may be necessary to keep the frame properly aligned while the erection sequence is in progress. Temporary supports may also be necessary until certain parts are properly secured by bolting or welding. The proper scheduling of required heavy equipment to be at the site would also be a concern of the erector. On large-scale projects, equipment or special devices necessary for the erection process may require heavier connections than those shown on the structural drawings. An erection scheme before the process begins would predetermine the location and weight of cranes and other equipment.

When the fabricated pieces of steel arrive at the job site they are placed by a steel erection crew in accordance with an *erection plan,* which is prepared by the fabricator. The erection plans are, in a sense, similar to the structural framing plans. The erection plan is a two-dimensional line drawing showing the framing at each floor, and where the sides of the building involve more than simply columns, a line drawing of the building frame elevation would be included. A sample of an erection plan is shown on Plate 4. The erection plans indicate the location of each piece of steel

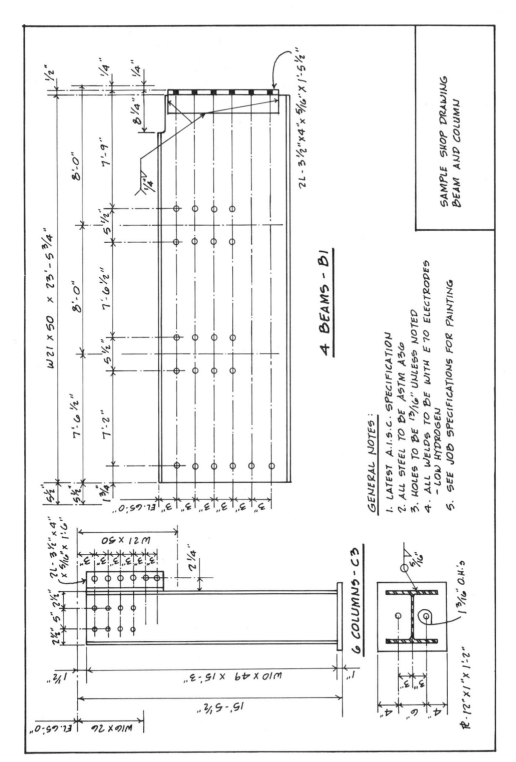

Plate 3 Sample of fabricator shop drawing.

17

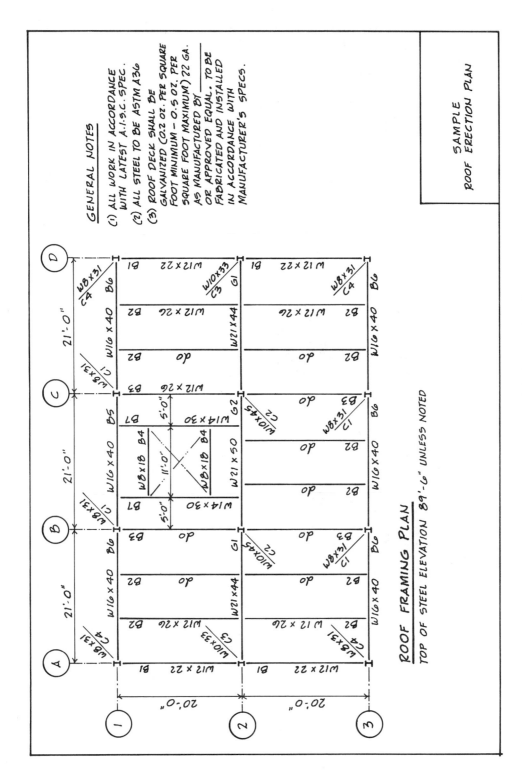

Plate 4 Sample of fabricator erection plan.

to be placed, and each piece is marked with a letter or number or a combination of both. The same marks are painted on the pieces at the fabrication plant. The orientation of the piece, if it can only fit one way, must also be indicated on the erection plan. For this purpose the piece mark is painted near one end and is shown similarly on the erection plan. The marking techniques used vary from one fabricator to another. In any case, clarity and coordination are the key words for a smooth and uneventful completion of the erection of the steel frame. Obviously, errors on the erection plans can lead to a great deal of confusion and time delays. Consequently, the erection plans, as are all of the shop drawings, are subjected to a great deal of scrutiny before the pieces are fabricated and shipped.

The following is an outline of the step-by-step procedures that are normally followed.

1. Bidding
   (a) Architect and engineer provide contract documents:
      (1) Specifications
      (2) Architectural drawings
      (3) Structural drawings
   (b) The fabricator determines the amount of material required, including shapes, sizes, lengths, and work to be done. The fabricator determines the cost for the project and submits a bid.
2. Postaward stage
   (a) The fabricator who is awarded a contract is provided with architectural and engineering drawings that are approved for construction.
   (b) The fabricator develops a material list itemizing the steel requirements for a particular job. The list is arranged into similar structural shapes and plates to order material from the mills in the most efficient manner. The fabricator also considers possible extra costs which may be based on:
      (1) Shape
      (2) Length
      (3) Quantity
      (4) Grade of steel
   (c) Development of connection details
      (1) The fabricator's engineering department isolates all connections that are not standard connections or are not fully detailed on the structural engineer's drawings. (We discuss this a bit more in Chapter 6.)
      (2) The fabricator's engineer sizes connection material, for nonstandard connections, in accordance with the loads shown on the structural drawings. This includes general configuration, size of plates and angles, number and size of bolts, and length and location of welds.
      (3) Connection details are submitted to the structural engineer for approval before shop detailing has substantially started.

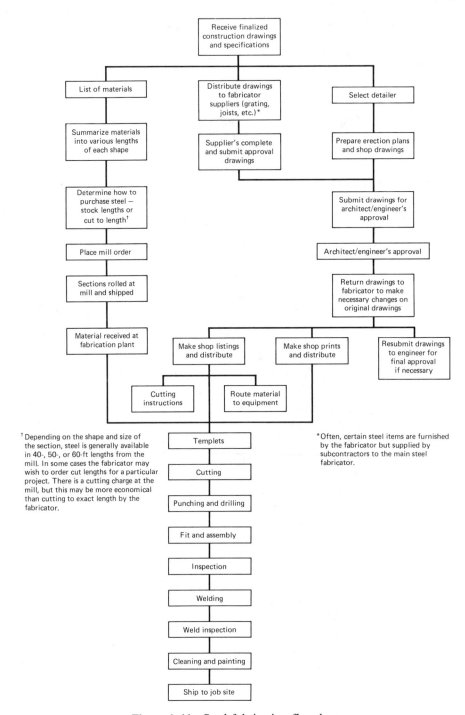

Figure 1-11   Steel fabrication flowchart.

    (d) Preparation of the shop drawings
        (1) The fabricator's detailer draws a picture for each piece of steel to be fabricated showing the specific work to be performed. The drawings show general configuration, hole locations, plates, connection angles, bolts, copes, weld sizes, and so on.
        (2) The fabricator checks shop details, particularly for general fit and dimensions.
  3. Architect's and/or structural engineer's approval
    (a) The fabricator submits shop drawings to the structural engineer for approval.
    (b) Shop drawings are reviewed for:
        (1) Correct interpretation of structural drawings
        (2) Correct size of supporting members
        (3) Correct number of bolts in connections (shop and field)
        (4) Correct amount of weld
    (c) The structural engineer returns shop drawings to the fabricator with comments and status, as follows.
        (1) Approved as submitted—proceed.
        (2) Approved subject to comments—proceed.
        (3) Rejected—corrections must be made and shop drawings resubmitted.
  4. Erection plans
    (a) Description—shows the steel erector where each piece is to be installed and the field welding required.
    (b) Erection plans are developed simultaneously with the shop details. The location and the number of the piece is immediately recorded on the erection plan.
    (c) The erection plans are submitted to the structural engineer in various forms of completion. It is the only document that correlates shop drawings to field location (and consequently, the location shown on structural drawings).

As can be seen from the outline of procedures, a number of checks are made, all in the interest of producing a final project where errors or confusion can be quite costly. The procedures outlined are in the interest of an uneventful project which is to the benefit of all involved. In this light, the steel fabricator can often be the structural engineer's "best friend."

The fabricator must also follow certain procedures internally, during the fabrication process. Figure 1–11 is a flowchart showing the events that take place within the fabrication plant and the interrelationship between these events.

SUPPLEMENTARY REFERENCES

Architectural Review, Vol. 72, November 1932 (entire issue is devoted to the history of steel and concrete.)

BIRKMIRE, WILLIAM H. *Skeleton Construction in Buildings,* 2nd ed. New York: John Wiley & Sons, Inc., 1894. Reprint edition by Arno Press, New York, 1972.

CONDIT, CARL W. *The Rise of the Skyscraper.* Chicago: The University of Chicago Press, 1952.

CONDIT, CARL W. *American Building,* 2nd ed. Chapter 10, "The Chicago Achievement: Steel Framed Construction"; Chapter 15, "The Steel Frame: Skyscrapers and Railroad Terminals"; Chapter 16, "New Departures in Steel Framing." Chicago: The University of Chicago Press, 1982.

COWAN, HENRY J. *The Master Builders.* New York: John Wiley & Sons, Inc., 1977.

COWAN, HENRY J. *Science and Building.* New York: John Wiley & Sons, Inc., 1978.

GALAMBOS, THEODORE V. "History of Steel Beam Design," *AISC Engineering Journal,* Vol. 14, No. 4, Fourth Quarter 1977, pp. 144–147.

GIDEON, SIGFRIED. *Space, Time and Architecture,* 5th ed. Cambridge, Mass.: Harvard University Press, 1967, pp. 167–208.

HART, F., HENN, W., and SONTAG, H. *Multi-story Buildings in Steel.* G. Bernard Godfrey, Editor of English edition. New York: John Wiley & Sons, Inc., 1978.

KIHLSTEDT, FOLKE T. "The Crystal Palace," *Scientific American,* Vol. 251, No. 4, October 1984, pp. 132–143.

KIRBY, RICHARD S., et al. *Engineering in History.* Chapter 10, "Iron and Steel." New York: McGraw-Hill Book Company, 1956.

ROWE, COLIN. "Chicago Frame," *Architectural Review,* Vol. 120, November 1956, pp. 285–289.

Fabrication and Erection

Engineering for Steel Construction. Chicago: American Institute of Steel Construction, 1984.

RAPP, WILLIAM G. *Construction of Structural Steel Building Frames.* New York: John Wiley & Sons, Inc., 1968.

2

FORCES ACTING
ON BUILDING STRUCTURES

For our purposes, a force may be defined as an action on a structural member which tends to change the state of motion of the member. Obviously, in an architectural structure, members must be designed so that they will remain in a state of *rest* when acted on by the variety of possible forces to which a building may be subjected. These include forces due to gravity, wind, earthquake, impact, vibration, temperature change (causing expansion or contraction), snow, ice, pressure due to liquids or earth, and so on. The probable intensity of some of these forces varies regionally.

Information regarding the intensity of loads to be used for structural design may be found in building codes. The recommendations for design loads given in these codes will vary, to some degree. Consequently, the appropriate building code for a particular locale should be used to determine the recommended intensity of forces to be used.

In addition to building codes, there are industry design specifications that are often adopted by building codes. In most areas (certainly, urban areas) laws require that applicable codes be followed. If an industry specification is adopted by the code, it, too, is part of the law. The industry specification to which we will make reference throughout this book is the specification of the American Institute of Steel Construction (AISC Specification).

GRAVITY LOADS

Steel-framed structures, as all building structures, must be designed to resist a variety of forces to which they will be subjected during their lifetimes. The primary forces, although not necessarily the most intense, to which a building is subjected are the forces produced by gravity. In building design, gravity forces are broken into two categories. They are as follows:

1. *Dead load.* This includes the weight of all permanent parts of the building, such as the structural frame itself, permanent partitions, ceilings, ductwork, lighting fixtures, and so on. The dead load of a building may be calculated with some precision based on the knowledge of the weights of various building materials. The Appendix to this book contains a list of the more common building materials and their weights. While the dead load can, in fact, be calculated precisely, such precision can be attained only after the structural frame has been designed. Therefore, during the structural design process the estimation of dead loads is, to some degree, based on sound judgment and experience.

2. *Live load.* The live load that must be supported by a structure includes the weight of all items that are not necessarily permanent, such as people, furniture, and so on. Based on this definition, it should be recognized that the live load cannot be calculated with any degree of precision since the presence, absence, or arrangement of nonpermanent items is unpredictable. In fact, the live load can only be estimated based on the function of the space. Live-load recommendations for a variety of spatial functions may be found in building codes. A sampling of such recommendations is given in Table 2-1. These recommendations are based, strictly, on experience and sound judgment. Live-load recommendations may vary slightly from one building code to another, and the governing code should be consulted.

It should be pointed out that live-load recommendations generally tend to be quite conservative. To illustrate this, try a little experiment. If you are in a classroom, measure the size of the classroom and determine the number of square feet of floor area. Then estimate the weight of your classmates, chairs, tables, and other nonpermanent items. Divide the total weight by the number of square feet and you will find that on a "per square foot" basis the actual live load will be very small. Compare this actual live load to the normally recommended design live load for classrooms of 40 lb/ft^2. You will quickly realize that code recommendations are generally based on a "worst possible case" attitude. Try the same experiment when you go home, by estimating the weight, for example, of all the items in your living room and dividing by the floor area. Residential spaces commonly carry a recommended design live load of 40 lb/ft^2. Again, you will find that the actual live load is not even close to the recommended value. Think about what it would take to develop the full recommended live load in such spaces and you will find that it would be unrealistic, under normal circumstances, for this to happen. One possible exception to this, in a residential structure, would be in an area supporting a water bed. Since water weighs 62.4 lb/ft^3, the 40-lb/ft^2 recommendation could be exceeded in the local supporting area; so be careful! In spite of what would normally be an excessive recommended design live load, such recommendations should be strictly followed because of the possibility that sometime during the life of a structure, an unusual situation may occur which would cause the recommended live load to be realized.

Based on probability, building codes generally recognize the fact that for

TABLE 2–1   LIVE-LOAD RECOMMENDATIONS

Function	Live load (lb/ft^2)	Function	Live load (lb/ft^2)
Assembly		Industrial	
Assembly halls, auditoriums,		Manufacturing	
churches, etc.		Light	125
Fixed seats	60	Heavy	250
Movable	100	Laboratories	100
Restaurants, gymnasiums,		Institutional	
grandstands, etc.	100	Hospitals	
Theaters		Wards and private rooms	40
Aisles and lobbies	100	Operating rooms	60
Balconies	60	Corridors	80
Stage floors	150	Residential	
Business		Private dwellings	
Offices	50	First floor	40
File rooms		Upper floors	30
Letter files[a]	80	Uninhabitable attics	20
Card files[a]	125	Multifamily	
Computer rooms	Review	Apartments	40
Lobbies and corridors	100	Corridors	80
Educational		Hotels	
School buildings		Guest rooms	40
Classrooms	40	Public rooms	100
Corridors	100	Corridors	100
Libraries			
Reading rooms	60		
Stacks	150		

[a]Depending on anticipated density of file cabinets, the values indicated can easily be exceeded. This situation should be reviewed as part of the design process.

members supporting relatively large floor areas, the full recommended live load will never have to be carried. Consequently, building codes will, with certain restrictions, allow some reduction in the recommended design live load, for members supporting large areas. Such allowances are generally based on the recommended live load to be carried and the floor area supported by a member. To take advantage of allowable reductions to the live loads, the governing building code in your area should be consulted.

Although not specifically under the heading of anticipated or calculable loads, such as live and dead loads, *construction loads* are something of which the architect or architect's representative on a building project should be keenly aware. Such loadings may occur during the construction phase of a building, where materials to be installed are stacked in a concentrated location. This can cause very intense loads which may cause some supporting members to be stressed beyond their elastic limit, or even worse, a failure. Even though such loadings are short-term loads, serious problems can result. Not only can this situation occur during the construction phase of a new building, but caution should be exercised in cases where remodeling or rehabilitation of an existing building is in progress. The supervisory procedures during the execution of a building project must include this concern. Materials to be used for the project should be spread out as much as possible to avoid intense loading on any particular member.

In addition to dead and live loads, which are the most prevalent of the gravity loads, there are other loads that must be considered in the design of a structure. For example, *snow loads* can be very intense, depending on the geographical location in which a building is to be built. Figure 2–1 is a generalized snow-load map of the adjacent 48 United States. This map shows the load, in pounds per square foot, of the horizontal projection of the roof surface, which may be expected in the various regions of the United States. However, it must be realized that many variations are possible since the manner in which snow may accumulate is unpredictable. For example, the values shown on the map may vary considerably in higher elevations. This may occur in various areas in the west and the northeastern parts of the country. The local governing building code should always be consulted. Also, the geometry of the building roof may have an effect on the snow load to be used. Blowing winds, during and after a snowstorm, may cause drifting and, consequently, heavy loads on certain parts of a roof, especially where the roof has valleys. In all cases, knowledge of the history of snow loading in various areas, which, presumably, is the basis for code recommendations, can be most useful.

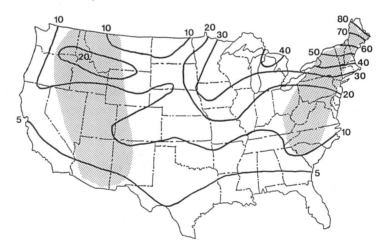

Figure 2–1   Generalized snow-load map. The shaded areas indicate regions where extreme variations in snow loading may occur. Local codes should be consulted and knowledge of local snow history would be valuable in making a final judgment.

Generally, snow may be thought of as weighing 0.5 lb/ft^2 per inch of thickness for a very dry snow and increasing for a dense wet snow which, possibly, can weigh as much as about 5 lb/ft^2 per inch of thickness. The values shown on the map will gradually diminish to the south of the contours.

Impact loading may be of concern where machinery is being supported, grandstands, and so on. Impact loads can be rather intense, and their effect on supporting structural members must be considered. The AISC Specification recommends that the live load for beams supporting elevators be increased by 100% to account for the effect of impact. A 33% increase in the live load is recommended for hangers supporting floors or balconies. Machinery loads should be increased from 20% to 50% depending on whether it is light machinery or reciprocating machinery.

LATERAL LOADS

Lateral loads affect the design of structures in all geographical locations. Lateral forces may be due to wind or earthquake. It should be mentioned that although both wind and earthquake forces subject a building to, essentially, lateral forces, the behavior of a building structure subject to such forces is quite different. The severity of wind forces acting on a building is a function of the exposed surface, while the effects of earthquake forces are based on the building mass. Although perhaps somewhat simplistic, it may be said that wind forces affect a building from the "top down," and earthquake forces (since they are introduced at the ground level) affect the building from the "bottom up."

Wind Forces

Figure 2–2 shows a generalized wind-loading map of the adjacent 48 United States, and the potential wind velocities (in miles per hour) are given for the various geographical regions. Approximate static pressures produced by these wind velocities, in terms of pounds per square foot, acting on the surface of a building, may be determed by

$$P = 0.003V^2$$

where P = pressure, pounds per square foot
   V = wind velocity, miles per hour

Actually, depending on a variety of conditions, the equation given above may be an oversimplification and is not intended for final design purposes. The determination of wind forces to be used for a building can be fairly complex, as there are several factors to be considered in addition to geographic location and wind velocities. For example, there will not only be pressures on the windward side of a building, but there will also be suction forces on the leeward side of the building. The intensity of the pressure and suction is largely a function of the shape and height of the building. Building codes provide recommendations for *shape coefficients* to be used to alter the basic equation. These coefficients are based on the proportions and geometry

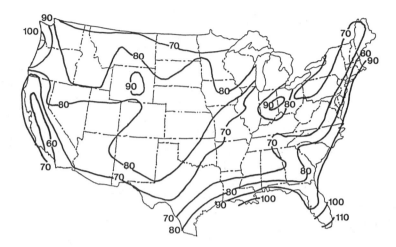

Figure 2-2 Wind-velocity map.

of the building. Also, building codes will provide recommendations that take into account the effects of the surrounding terrain and other physical features.

Not only will wind produce pressures and suctions on exposed vertical surfaces, but depending on the shape of a roof, there may be pressure or suction perpendicular to the roof plane. For example, if a roof is flat, there will be suction forces acting perpendicular to the surface. For a gable roof, if the slope is greater than about 30°, there will be pressure on the windward side and suction of the leeward side, as shown in Figure 2-3. Again, appropriate building codes should be used as a reference for determining values.

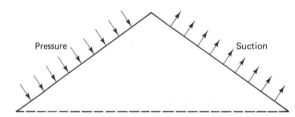

Pressure   Suction

Figure 2-3 Wind forces on gable roof; slope of roof greater than 30°.

For very tall buildings the effects of wind forces are extraordinarily complex and, generally, the effects of wind and the structural design must be based on wind-tunnel analysis of a model. In every case, the governing building code should be used as a reference to determine the magnitude of the forces due to wind for which a building should be designed.

Seismic Forces

Figure 2-4 shows a generalized seismic risk map of the United States, indicating the severity of earthquakes that may occur in various geographic locations. Earthquake forces are extraordinarily intense, and, to a large degree, their effect on a building

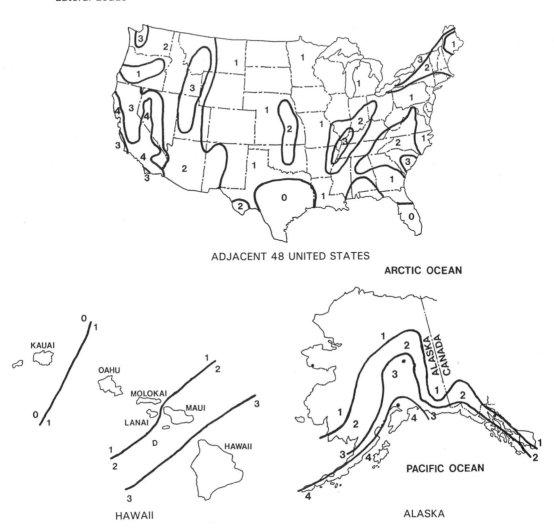

ADJACENT 48 UNITED STATES

HAWAII

ALASKA

Figure 2-4 Generalized seismic risk map: zone 0, no damage; zone 1, minor damage; zone 2, moderate damage; zone 3, major damage; zone 4, potential for extremely major damage.

structure is unpredictable. A great deal of research is constantly under way to improve structural design procedures with respect to earthquake forces. Steel-framed structures provide a building with the very desirable property of ductility. Although it may sound contradictory, it is also desirable that the building subjected to seismic forces be reasonably stiff. It is the combination of ductility, stiffness, and rigidity that makes the design of a building to resist earthquake forces terribly complex.

"As far as the structural engineer is concerned, the most important aspect of earthquake engineering is this basic difference from all other structural design. Our design forces are only a *small* fraction of the forces expected to be exerted on a structure in a major earthquake. In other words, the structure *will* be overstressed many times as defined by usual design standards.

The structure must remain coherent and stable at deformations of many times the yield deflection. This observation, in turn, means that in designing to resist earthquake forces, we not only have to consider specific forces and loads i.e., provide certain minimum *strengths* (as in most structural engineering design), but we must also consider the performance at great overloads and large deformations. This affects joint and member detailing to assure that the structure will hang together at large deformations and affects member proportions so that less critical elements fail first and absorb energy and so help to protect the more critical members such as columns. This must be emphasized over and over again: *In earthquake resistant design, it is not sufficient or adequate to make a member "strong enough"; it must also have a reserve of ductility. The whole concept of structural design changes.*"*

DESIGN METHODS

There are two methods by which structural steel members may be designed. They are the allowable stress design method and the plastic design method. Let us briefly define the two methods and we will elaborate in the chapters that deal, specifically, with these procedures (Chapters 4, 5, and 8).

1. *Allowable stress design.* In this method the design procedures are based on recommended allowable stresses for the grade of steel being used. The allowable stress is well within the elastic limit of the material. In this procedure the members are designed based on the actual dead load, live load, and lateral forces that must be resisted. This method has also been known as the *elastic design method.* It may be recalled, from basic studies, that an elastic analysis of statically indeterminate members (continuous beams and frames) involves a great deal of tedious computation. Of course, the availability of computers has removed much of the tedium that would be encountered in the solution of simultaneous equations, using the classic Theorem of Three Moments and the procedures involved in frame analysis.

2. *Plastic design.* This method is one that takes into account the ductility of structural steel and the great reserve strength of this material. Unlike the allowable stress design method, in the plastic design method the members are designed so that the full cross section, at the points of maximum moment, will be at the yield stress of the material. Essentially, plastic design is a method in which, philosophically, the design of the structural member is based on the maximum load that may be supported until "failure" occurs. While the yield stress of the material is used in the design process, the loads used are based on the anticipated loads which are hypothetically inflated and then viewed as ultimate loads. It is under these ultimate loads, which will not be realized in the structure, that the member is designed to "fail."

*Henry J. Degenkolb, *Earthquake Forces on Tall Structures,* Bethlehem Steel Corporation, Booklet 2717A, 1977. See this reference for further discussion on this matter.

As will be seen in Chapter 8, this philosophy of design makes the computational aspects of indeterminate structures much simpler than in the allowable stress design method. Plastic design serves little or no purpose in the design of simply supported beams. In addition to simpler computational procedures for continuous beams and frames, plastic design procedures will generally indicate smaller member sizes compared to the requirements indicated by elastic methods.

FACTORS OF SAFETY AND FAILURES

In steel-framed structures, as in all structures, we have factors of safety included in the process of design. Factors of safety may be rather large in the design considerations, but they are quite necessary. Essentially, a factor of safety may be defined as the ratio of the ultimate strength of a structural member to its maximum anticipated stress. An appropriate question would be: Why have large factors of safety? There are a variety of reasons for this. Some of the reasons are as follows:

1. The actual strength of a member may change with time due to corrosion, fatigue, and so on. Although steel, in most structures, is generally protected from corrosion by either fireproofing material or painting, there can be no guarantee that appropriate maintenance will be provided as needed.

2. The analysis of a structure may have some errors. Although this is a possibility, it should be recognized that, normally, a structural design goes through a variety of checks before the actual structure is built. This tends to minimize, certainly, any serious errors in the design procedures.

3. Subjection to unpredictable forces and magnitudes is a basis for some concern. The process of design that is normally followed by building design professionals includes strict adherence to building code requirements, which, based on experience and observation, tends to take much of the unpredictability out of the nature of forces and their magnitudes. However, on occasion, such as in a severe earthquake, magnitudes of forces may be reached which could not realistically be predicted or anticipated.

4. Residual stresses due to the fabrication and erection process may be very high. Residual stresses may be due to a variety of situations that occur during the milling, shipping, and fabrication process. These may be due to members being distorted during the shipping process, the punching of holes, welding of items, and other operations that normally take place during the fabrication and erection process. The stresses created by these processes may, indeed, be very high before the load is even placed on the structural member.

5. Construction loads that occur during the process of erection, although normally monitored, may still be quite unpredictable. The stacking of materials in a concentrated area may cause stresses that are well beyond the allowable stress of the material, albeit for a short-term period.

The key word in all of the above is "unpredictability." Although we have very sound engineering practices, building codes, and a variety of available material to research that seem to guide us precisely in the design of a steel structure, reality suggests that none of these can be regarded as absolute. Therefore, factors of safety that are recommended are not only necessary, but they must be considered as welcome guidelines to the designers of a steel-framed structure.

In spite of all the good engineering practices, well-documented building codes, educated judgment, and factors of safety, failures occasionally occur in structures. Because of the very sound industry standards provided, the few failures that do occur normally do so during the erection process of a building and the problem is generally a fault of the process and not of the material. There are a variety of reasons for this, but, in general, it may be said that erection failures mostly occur during the process of construction simply because the various parts of a building have not yet been tied together and do not yet have their anticipated stability since, at this point, they are only two-dimensional configurations. Individual frames that have not yet been properly braced are simply not very stable. It may, in fact, be said that during the erection process we have parts acting independently which are not yet part of a building. It is primarily during this phase of construction, when the structural system is not tied together, that extreme care must be exercised to ensure stability during the interim period.

Although the literal collapse of a completed structure is, fortunately, a very rare occurrence, there are, nevertheless, "failures" that are not uncommon. This depends on how one wishes to define "failure." In an architectural structure there may be many situations that one might view as constituting failure in the sense that the building may not be properly serving some of its intended functions. For example, foundation problems may occur and if settlements of the building are widely varied, this will cause distortion to the supporting structural frame with the consequence of additional stresses well beyond those initially anticipated (see Figure 2-5). Although the building may not collapse in a situation such as this, there will probably be serious damage to architectural finishes due to the distorted frame. The

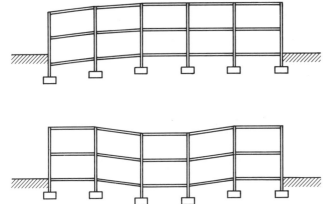

Figure 2-5 Frame distortions due to differential settlement.

problems may also be compounded by leakage, broken glass, and pieces of material falling. Although there may be no threat to life, it may be said that such conditions represent a building failure.

In addition to this problem, members that deflect excessively may also cause an "architectural failure." For example, if a beam is spanning over a large glass pane and excessive deflection occurs, the beam will try to rest on the glass, causing the glass to bow or break. An excessively deflected beam supporting an exterior masonry wall may result in severe cracking of the wall, and the problem may then be compounded by water penetration, causing further damage to interior finishes. Again, there may be no threat to life, but the usefulness of the building, or part of it, may be in question. The architect, especially, must be keenly aware of the manner in which the structural frame and the architectural components interact. Good judgment in detailing procedures, accuracy in design, and sensitivity toward possibilities of problems that may occur because of design decisions will tend to minimize, if not eliminate, "failures."

SUPPLEMENTARY REFERENCES

Architects and Earthquakes. Washington, D.C.: AIA Research Corporation, 1975.

A.S.C.E. Task Committee on Wind Forces. "Wind Forces on Structures (Final Report)," *Transactions of the American Society of Civil Engineers,* Vol. 126, Part 2, 1961, pp. 1124–1198.

BIGGS, JOHN M. "Wind Forces on Structures: Introduction and History," *Journal of the American Society of Civil Engineers,* Vol. 84, St. 4, Part 1, Paper 1707, July 1958.

DUKE, MARTIN C. "Points to Consider: Quakes and Tall Buildings," *AIA Journal,* Vol. 59, No. 1, January 1973, pp. 36–37.

KOEHLER, ROBERT E. "Comment and Opinion: The Trembling Truth about Earthquakes," *AIA Journal,* Vol. 58, No. 5, November 1972, p. 4.

SNOW, JOHN T. "The Tornado," *Scientific American,* Vol. 250, No. 4, April 1984, pp. 86–96.

SZENDY, E. J. "Wind Load, Relating Speed to Pressure," *Architectural and Engineering News,* April 1966, pp. 50–65.

WOODRUFF, GLEN B., and KOZAK, JOHN J. "Wind Forces on Structures: Fundamental Considerations," *Journal of the American Society of Civil Engineers,* Structural Division, Vol. 84, St. 4, Part 1, Paper 1709, July 1958.

Major Building Codes

BOCA Basic Building Code. Building Officials and Code Administrators International.

Building Code Requirements for Minimum Design Loads in Buildings and Other Structures, ANSI A58.1–1982. New York: American National Standards Institute, 1982.

New York City Building Code.

Standard Building Code. Southern Building Code Congress.

Uniform Building Code. International Conference of Building Officials.

3

SYSTEMS IN STRUCTURAL STEEL

THE SKELETAL FRAME

The most common of steel-framed structures is the skeletal structure with a rectangular bay arrangement, as shown in Figure 3–1. This sort of framing is often referred to as *beam and column construction*. This system of framing has been utilized effectively for buildings of all sizes and heights. For tall buildings, wind and seismic forces must be considered and the frame must be braced properly. We discuss bracing systems and lateral forces in Chapter 7. In this chapter we concentrate on the configurational aspects of steel framing.

The determination of the bay dimensions, in a beam and column structure, can be a relatively complex procedure. Indeed, there is no simple formula to assist the architectural designer in this decision. A variety of factors must be evaluated to determine the appropriate bay dimensions for a given situation. The primary determinant, it seems, would be the function of the building and the arrangement of internal spaces. This would include the necessary location of partitions as well as the dimensions of the partitioning materials. Also to be considered is the spacing of columns to accommodate the exterior wall materials.

Closely coordinated with the column spacing decision must, of course, be the economy of the structural steel frame itself and the floor system to be supported. In general, reasonable economy can be achieved with column spacings somewhere on the order of 20 to 30 ft. However, larger spacings are often used where the demands of the internal spaces make this feasible. It should be noted that long clear spans require larger amounts of steel and the tonnage required will be dramatically increased compared to normal spans. Also, when dealing with longer-than-normal spans, deflec-

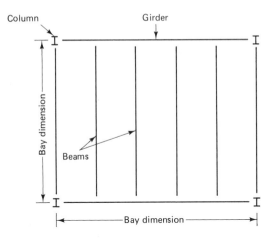

Figure 3-1 Plan view—typical rectangular bay.

tions may become a critical issue and should be evaluated. The techniques for evaluating deflections for a variety of situations are presented in Chapter 4.

In addition to the determination of bay dimensions, which will dictate the clear span for the girders in a rectangular bay system, consideration must also be given to the beam spacing within the bay. The beams, which are supported by the girders, must in turn support the floor system. In lieu of steel wide-flange beams, open-web steel joists may be used. Steel joists are, in terms of behavior, nothing more than beams. A typical open-web steel joist is shown in Figure 3-2.* They are often referred to as *bar joists* because the web and bottom chord are generally made from bar stock. Standard joists are normally spaced closely, and the floor deck is therefore very light. Because of the open web, ductwork and other utilities can easily be placed within the depth of the member.

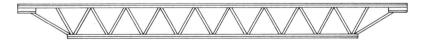

Figure 3-2 Open-web bar joist.

There are a large variety of floor decks that may be suitable for use with a steel frame structure. The spanning capabilities of these floor decks vary, depending on a variety of factors. The most commonly used material is the steel deck. The steel deck is corrugated, or folded, to give it strength. The deck is topped with concrete, which creates a stiff floor. The steel deck is made from sheet steel, which is available in a variety of gauges. The depth of the corrugations, or folds, and the spacing of the folds are also variables, as is the thickness of concrete and the reinforcing. Some typical sections of metal deck systems are shown in Figure 3-3. There are a number of manufacturers of metal decks and there are a variety of depths and metal gauges

*See *Sweet's Catalog File* for manufacturers and variety available.

Figure 3-3 Typical metal deck profiles.

available to satisfy a large number of span and load-carrying requirements.* However, it must be noted that the decks made for long spans are generally rather costly compared to the lighter-gauge and shallower decks made for shorter spans.

In addition to metal deck systems, it is also possible to use precast concrete planks. These are also available in a variety of depths and spanning capabilities. Several typical sectons of precast concrete floor decks are shown in Figure 3-4. Normally, precast planks are also prestressed.†

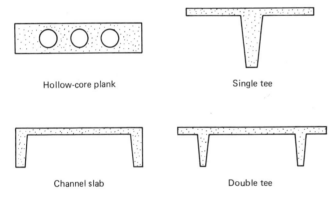

Hollow-core plank

Single tee

Channel slab

Double tee

Figure 3-4 Some typical precast concrete floor decks.

Finally, it is also possible to use cast-in-place concrete to span between beams. When this system is used, a one-way slab system is the most commonly used, although it is possible to use a two-way concrete slab. A two-way slab is advantageous because intermediate beams within the bay are eliminated. Concrete floor systems, whether precast or cast-in-place, have the additional advantage of being inherently fireproof. Consequently, only the steel frame needs to be fireproofed. When metal deck is used, the underside of the deck must be fireproofed (in buildings where fire resistance is required) unless a ceiling is used which is capable of providing fire resistance.

Because of cost, the cast-in-place concrete floor system is not particularly common in a steel-framed structure. The inherent advantages are often offset by the need

*For those readers who wish to see the variety of deck systems, reference should be made to the *Sweet's Catalog File,* which is a collection of published data furnished by manufacturers of building products.

†Prestressing is, basically, a technique that produces residual compressive stresses in a concrete member. These residual stresses are intended to negate tensile stress that would be produced by the building loads. Prestressing also provides an extra measure of stiffness by controlling deflections.

for formwork and the delays involved in waiting for the concrete to attain some strength before the forms and shoring are removed. The steel deck with concrete topping seems to be the most popular system for steel-framed buildings. The deck serves as a form for the concrete, and a working surface is provided almost immediately. This speeds up the construction process because other trades can proceed with their work as soon as the concrete has some strength, without waiting for shoring to be removed. Indeed, partitioning and other finish work can proceed on a lower level while floor decks are being installed above.

Steel decks may be either composite or noncomposite decks. That is, the poured concrete, depending on the deck design, may be considered to act in unison with the deck, or the deck may simply be a form for the concrete. In turn, whether a deck is composite or noncomposite, composite action may be achieved between the concrete and the supporting beams through the use of shear studs welded to the beam. The principles of composite design are discussed in more detail in Chapter 4. When a steel deck is used and composite action is desired between the concrete and the supporting beams, the shear studs can be welded directly through the deck, in most cases. *

Nonrectangular Arrangement

Although it is rarely seen, it is possible to build a skeleton framework of structural steel where the columns are spaced in a nonrectangular pattern. For most building types where structural steel is the primary structural system, the common arrangement is, indeed, a rectangular bay pattern. This provides a great deal of flexibility where it is necessary to rearrange partitions from time to time. However, certain building types with fixed spatial requirements may be suitable for a nonrectangular system of supports, where the supports are columns or short lengths of bearing walls, as shown in Figure 3–5.

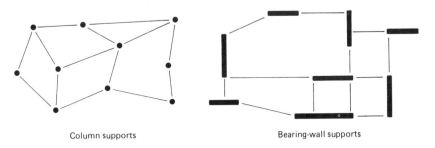

Column supports   Bearing-wall supports

Figure 3-5 Nonrectangular arrangement of supports.

For certain cases it may be quite interesting, architecturally, to design a nonrectangular system to control circulation patterns as well as spatial arrangements. Clearly, such arrangements would, largely, be limited to one-story buildings. For buildings of larger scale this can create a problem of economics. In a rectangular system, many

*See Commentary on the AISC Specification, Section 1.11.5, for restrictions on this procedure.

beams and girders are fabricated in precisely the same way and the connections are standard. With a nonrectangular column arrangement it is possible that each piece would have to be fabricated differently, and connections between members may be difficult. This, of course, increases cost, which must be considered in the design process.

BEARING WALLS

In addition to skeletal construction in structural steel, which is, certainly, the most common form of construction for buildings at least several stories in height, it is not uncommon to use bearing walls with steel beams for one- or, perhaps, two-story buildings. The same floor systems used for skeletal construction may also be used where the supports are bearing walls instead of columns. Bearing-wall construction with steel beams for the floor or roof supports are not particularly common for buildings over one or two stories for several reasons. Bearing walls, which would normally be made of masonry (i.e., brick or concrete block), tend to become exceedingly thick when supporting the load from many stories. Although the necessary thickness may be reduced by using reinforced masonry, this can be a costly procedure. In fact, masonry bearing walls must be reinforced in areas prone to earthquakes. In addition to this, there may be problems of timing and coordination since there are several trades involved. The masonry walls must be in place before the steel beams can be installed. The problems of coordination are minimal in low-rise buildings.

LONGSPAN STRUCTURES

There are many building types where it is desirable, if not necessary, to span long distances between supports in order to keep the space free from interference by columns. These include such buildings as gymnasiums, aircraft hangers, convention centers, covered stadiums, and so on. The clear spans required may vary from moderate to extreme. Obviously, the methods of construction and the structural systems used are quite different from those discussed previously. There are a variety of systems in structural steel that are commonly used where longspan requirements must be met. We will now discuss some of these systems.

Trusses

Essentially, a truss is a deep beam with an open web. It is made of small pieces arranged in triangular patterns, as shown in Figure 3–6. Steel trusses are, perhaps, one of the most common types of structures used for long clear spans. Because of the arrangement of the pieces in a truss, long spans may be achieved with minimum amounts of material. Steel trusses can be used effectively for spans of several hundred feet.

While it is assumed that the reader has been exposed to the concepts and nomenclature of trussed structures, it seems that it will be useful here, as a refresher,

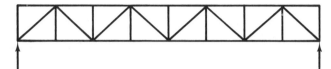

Figure 3-6 Truss.

to repeat some of the information regarding the nomenclature of a typical truss. As shown in Figure 3-7, trusses may be flat or pitched. The member on top of the truss, whether sloped or flat, is known as the *top chord.* Conversely, the member on the bottom of the truss is known as the *bottom chord.* The members that form the tri-angulated pattern between the top and bottom chords are known as the *web members,* whether they are diagonally oriented or vertical. The space between the points where web members meet the chords is known as a *panel* and the points where these come together are known as the *panel points.* There are an infinite variety of truss configura-tions that may be developed using the concept of triangulation. Trusses may be sym-metrical or unsymmetrical. Several possibilities are shown in Figure 3-8.

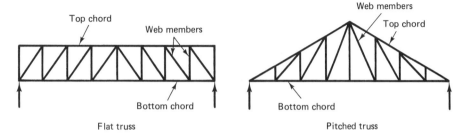

Figure 3-7 Truss nomenclature.

Warren truss.   The Warren truss is a flat top truss with web members arranged as shown in the figure. The vertical members may or may not be provided, depend-ing on a variety of factors, such as the span, depth, and distance between panel points. One of the more popular applications of the Warren truss configuration is the *longspan steel joist.* These are available from a variety of manufacturers and reference should be made to the various manufacturers' literature for load-carrying capacities and details. Longspan joists are available in depths up to 48 in. and for spans up to about 90 ft. Deep longspan joists are available in depths of up to 72 in. for spans of about 140 ft, for roofs.

For long spans on roofs it is recommended that some pitch be applied to the top chord in order to shed water properly to points where it can be drained. Gen-erally, the pitch is about $\frac{1}{8}$ in. per foot in order to achieve this goal. The pitch may rise to the center of the span or it may be to one end of the truss. In any case, even over a long span, this will visually appear to be a flat roof.

Flat Pratt truss.   This truss is configured as shown in Figure 3-8. The most economical range of spans is about 60 to 120 ft. It is generally recommended that some pitch be provided for the top chord in order to drain the roof. The Warren

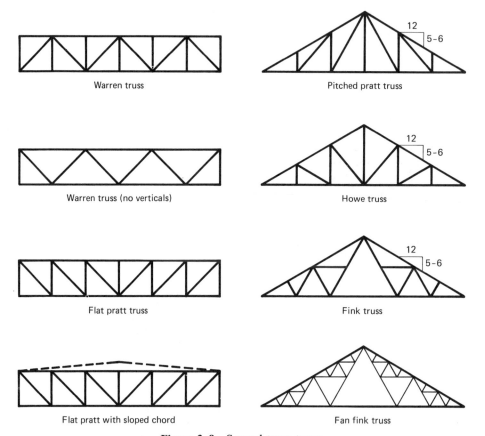

Figure 3-8   Several truss types.

truss is probably a bit more economical than the flat Pratt, especially where roof loads are light enough and a longspan joist would be suitable.

Pitched Pratt truss.   In addition to the flat Pratt, a Pratt truss may also be built with a sloped top chord. The slope generally used is shown in Figure 3-8. The economical range of spans is about 60 to 120 ft.

Howe truss.   This truss has web members that are sloped in the opposite direction from those in the pitched Pratt. The largest span, for economy, is also about 120 ft and the slope of the top chord that is generally used is the same as the pitched Pratt. The choice of using a Howe or pitched Pratt depends on a variety of issues. This includes the span, magnitude of load, and the load patterns that may be imposed on the truss. To determine which would be most economical would require a mathematical analysis.

Fink truss.   Several variations of the Fink truss are shown in Figure 3-8. This type of truss is economical for spans of about 100 to 125 ft when made of structural steel. Generally, the Fink truss may be more economical than the Howe or pitched Pratt, in terms of the amount of material required to do a given job. This is so because, in the Fink truss, the arrangement of the web members is such that most of the members are in a state of tension and those in a state of compression are relatively short. Tension members require less cross-sectional area to resist a force than do compression members of equal length. As the span in a Fink truss becomes large, it should be obvious from the figure that there are a number of small pieces involved and the economy in the amount of material can be more than offset by the cost of fabrication.

Truss Loading

As mentioned previously, there are a variety of truss configurations that may be developed and, indeed, where requirements are such that the truss is to be unsymmetrical, the designer would have to determine the arrangement of the parts based on a variety of conditions, such as load, span, decking system, and so on. Whether a designer is using one of the standard configurations, already discussed, or developing a unique arrangement for the conditions to be satisfied, the determination of the configuration and the economy of the truss is, to a great extent, dependent on the spacing between panel points. The spacing between panel points is largely a function of the roof deck that will be used.

It is imperative that all loads to be carried by a truss be delivered to the panel points. The truss is economical for long spans because, if properly designed and fabricated, all members will be stressed in a pure state of axial tension or compression. This is so because the truss is fabricated in a manner such that the centroidal axes of all members coming together at a panel point have a common point of intersection. This is achieved through appropriate detailing as shown in Figure 3–9. When

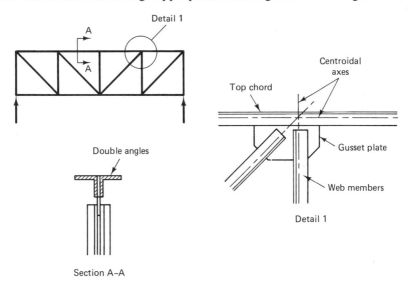

Figure 3–9 Truss details.

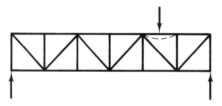

Figure 3-10 Bending in top chord if load is not delivered to the panel points.

the line of action of the superimposed load also intersects at this point, the members will be in a pure state of axial stress. If a load is delivered between panel points, as shown in Figure 3-10, bending will occur which would seriously reduce the economy and, indeed, violate the fundamental concept of appropriate truss behavior. It is, therefore, necessary that loads be delivered to the panel points. This is achieved by directing roof loads to these points with members called *purlins* (see Figure 3-11). Most often purlins are channel sections or wide-flange sections.

When used on a sloping surface the weak axis of the section is supported by sag rods which are tied together at the top of the slope (see Figure 3-11b). For nor-

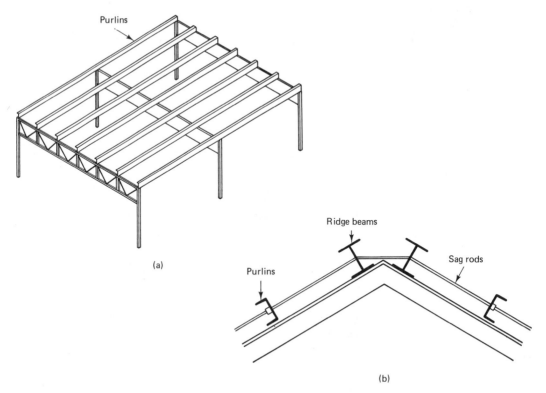

Figure 3-11 (a) Purlins delivery load to the panel points; (b) purlins on a sloping surface.

mal spacings between trusses usually one line of sag rods at the midspan of the pur-lins is necessary. The roof deck spans between the purlins, and since the purlins must fall at the panel points, consideration of the spanning ability of the roof deck must be an integral part of the truss design. The magnitude of loading delivered by the purlins is a function of the space between panel points and the spacing of the trusses. Generally, steel roof trusses may be effectively spaced about 16 to 24 ft apart. If large dimensions between panel points are used, the top chord, which will be in a state of compression under gravity loads, will become much larger than if the panel points are kept reasonably close. On the other hand, however, if the dimension be-tween panel points is small, there will be more web members involved. It should be obvious that to make an absolute statement about the best situation for economy would be an exercise in futility. There are a variety of considerations, all of which interact.

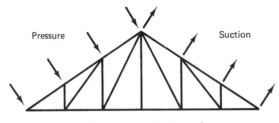

Pitched truss, greater than 30°

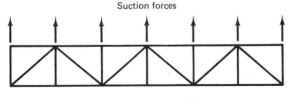

Flat truss, less than 30°

Figure 3–12 Effects of wind forces on trusses.

The analysis of a truss to determine the magnitude and character of stresses (i.e., tension or compression) is not presented here. Normally, the reader will have learned the procedure of analysis in a basic structures course.* It should be men-tioned here, however, that the analysis of a truss, especially with sloped top chords, must include consideration of wind forces. Depending on the slope of the chord it is possible that wind forces may produce pressure on the windward side and suction on the leeward side. Generally, this is so for roof slopes greater than 30°. Where the roof slope is less than 30°, including flat roofs, the wind is considered to produce negative pressure (suction). In all cases, whether the forces are pressures or suction, the wind forces are considered to act perpendicular to the surface, as shown in Figure 3–12. The appropriate building codes should be consulted where numerical data for

---

*For a refresher in the analytical procedures for trusses see I. Engel, *Structural Principles* (Englewood Cliffs, N.J.: Prentice-Hall, Inc., 1984).

wind forces is desired.* Depending on the slope of the top chord and the magnitude of wind forces, it is possible that an analysis would show reversal of the character of stress in certain members. Therefore, when analyzing the stresses in truss members several analyses should be made and the results superimposed to determine the worst conditions for individual members. For example, it is possible that a member acting in a state of tension under gravity loads will have its stress reversed under the action of wind loads. Although the magnitude of the compressive stress may be less than the tensile stress, a larger member may be required because of the problem of buckling in compression members. In other members the total stress may be increased due to wind acting with gravity loads and the member design would have to include these increased forces.

The AISC Specification recognizes the fact that stresses produced by wind or earthquake are short term in nature. Considering this, and the ductility of steel, the AISC provides an allowance for a one-third increase in the allowable stress in a member when produced by wind or seismic loading acting alone or in combination with gravity loads. The section determined using the increased allowable stress may be used provided that this section is not smaller than one required for gravity loads alone.

Three-Hinged Frames

In addition to trusses, there are other structural configurations that may be used for longspan requirements. One of these is known as a three-hinged frame. In this type of structure the spanning members are integral with the vertical supports, thereby producing continuity between the vertical and spanning members. A typical three-hinged frame is shown in Figure 3–13. The three-hinged frame is an efficient struc-

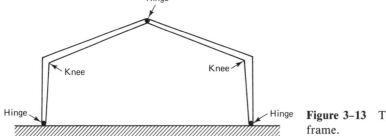

Figure 3–13 Typical three-hinged frame.

ture, capable of spanning long clear distances, and is easily analyzed using the basic principles of static equilibrium.

A *hinge,* as referred to in this context, has a meaning similar to a hinge that connects a door to a jamb. It may be said, quite simply, that a hinge is a device that connects two parts and allows rotation between them, while restraining translation.

---

*For a complete discussion on wind forces, an excellent reference is "Wind Forces on Structures," *Transactions of the American Society of Civil Engineers,* Vol. 126, Part 2, 1961, pp. 1124–1198.

In structural language this means that there can be no bending moment at a hinge, but the hinge will resist shearing forces.

The spanning capability of a three-hinged frame is made possible by the ability to resist large bending moments at the *knee* of the frame. Because of the moment-resisting requirements at the knee, the amount of material required at this point may be substantially greater than at other sections in either the verticals or spanning members. As mentioned previously, the three-hinged frame is easily analyzed by the equations of static equilibrium, and this is so because of the hinges, which are known points of zero bending moment.* Because of the continuity at the knee of the frame, there are, in addition to vertical reactions produced by the loading, also horizontal reactions required in order to keep the legs of the frame from moving outward, as shown in Figure 3–14. Consequently, the connection of the leg to the foundation must be appropriately designed and detailed in order to resist the outward thrust. The stress at this location will be a shearing stress. There will be no bending moment at this point because it is a hinge. There are a variety of ways to resist this thrust. One approach would simply be to provide bolts that are large enough to resist the shear.

To maintain the characteristics of a hinge, which would allow rotation between the leg and the foundation, the placement of bolts is important. To provide these characteristics the bolts should be placed to line up with the neutral axis of the member, as shown in Figure 3–15. This suggests that only two bolts should be used at the base.

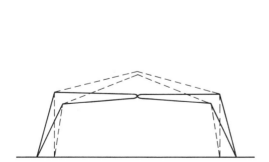

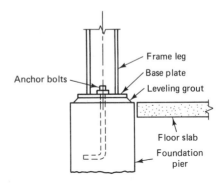

Figure 3–14 Legs of frame tend to move outward due to vertical load.

Figure 3–15 Base connection of frame.

Where shearing forces across the bolts are high, due to large horizontal thrusts, it may be necessary to provide a four-bolt connection, and since it is important that the bolts be placed in a symmetrical pattern, these would not line up with the neutral axis of the leg. Consequently, there would be some moment resistance at the base because of the lever arm between the bolts. If the distance between the bolts is kept small, the bending moment produced may be negligible.

*For the analytical procedures involved in three-hinged frames, see I. Engel, *Structural Principles* (Englewood Cliffs, N.J.: Prentice-Hall, Inc., 1984), pp. 126–132.

Another problem that may exist due to the outward thrust at the base of the legs is the tendency to cause the foundation to rotate, as shown in Figure 3–16. The most common approach used to deal with this issue is to use a tie-rod to connect the tops of the foundation piers or walls, as shown in Figure 3–17. The tie-rods will go into a state of tension when the legs of the frame try to move outward, thereby stabilizing the entire configuration. The tie-rod would be below the slab and, itself, encased in concrete.

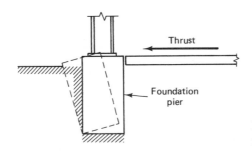

Figure 3–16 Foundation tends to rotate due to thrust.

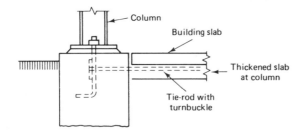

Figure 3–17 Tie-rod detail.

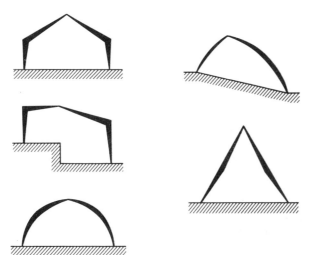

Figure 3–18 Several possibilities for three-hinged frame configurations.

Three-hinged frames, while normally symmetrical in configuration, need not be so, depending on the function being enclosed. The frames may be unsymmetrical and they may be based at different levels. Several possibilities for frame configuration are shown in Figure 3–18.

Loads are delivered to frames in much the same manner as they are to a truss. Purlins are used to support the roof deck, and the purlins deliver concentrated loads to the frame. Since there are no panel points, as there are in a truss, to which these loads are delivered, the loads will produce bending moments throughout the frame, except at the hinges. Generally, steel three-hinged frames are most effectively spaced at about 18 to 30 ft apart. Wind forces on three-hinged frames must be considered in precisely the same manner as discussed in the earlier section on trusses.

Two-Hinged Frames

In a two-hinged frame (most often referred to as a *rigid frame*) the hinge at the peak of the frame is eliminated by welding the two sections together (see Figure 3–19). Because of the absence of the third hinge, the two-hinged frame is a statically inde-terminate structure. This means that the analysis to determine the maximum moments is somewhat more complex than analysis of a three-hinged frame, which is a statical-ly determinate structure. Because of the complex computational procedures involved, this was, at one time, an important consideration in the overall economy of the struc-tural design. However, with the availability of computers and software today, this is no longer of any great concern. The two-hinged frame is, in terms of material usage, more economical and more efficient than a three-hinged frame. In a two-hinged frame the thrust at the base will be considerably less than for a three-hinged frame with the same load, span, and configuration. Also, in a two-hinged frame, the bending moment at the knee, which is normally the critical value for design, will be considerably less than that for a three-hinged frame. The same conditions must, however, be con-sidered when dealing with the thrust at the base of the legs. It is strongly recommended that, to avoid foundation difficulties, the base of the legs be tied together with a ten-sion tie-rod.

When designing a two-hinged frame the most economical approach is to use constant-section wide flanges, which must be appropriately detailed at the knee of the frame, where the internal forces are abruptly changing direction. However, as

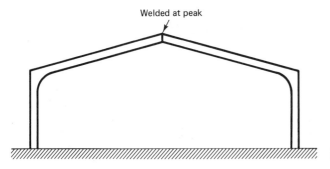

Figure 3–19 Typical rigid frame.

suggested previously, in some of the shapes shown in Figure 3–18, the frames may be fabricated from steel plates that are modeled in response to variations in the bending moment. Although this may result in a substantial savings in material compared to a constant section, the fabrication costs may be quite high. Whether using a constant-section wide flange or a specially fabricated frame, the knees of the frame may require a substantial addition of material in order to resist the large bending moment and the inherent behavior of stresses that occur in continuous members where the direction changes. Some possibilities for the detailing of the knee are shown in Figure 3–20. It should also be said that since such frames are normally exposed, the configuration of the knee may also be guided by aesthetic considerations.

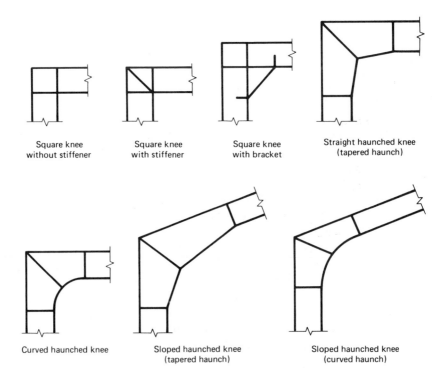

Square knee
without stiffener

Square knee
with stiffener

Square knee
with bracket

Straight haunched knee
(tapered haunch)

Curved haunched knee

Sloped haunched knee
(tapered haunch)

Sloped haunched knee
(curved haunch)

Figure 3–20 Knee details. From *Engineering for Steel Construction*, used with permission of The Americal Institute of Steel Construction.

Like three-hinged frames, the steel rigid frame is normally spaced at about 18 to 30 ft. However, spacings somewhat more or less are also possible. This decision will affect the size of the purlins required and the total load being delivered to the frame. In addition, rigid frames are quite efficient for spans of about 50 to 150 ft.

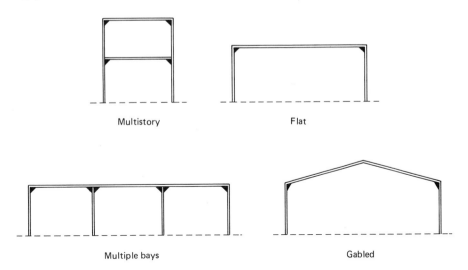

Figure 3-21   Rigid frame types.

They may be flat or gabled and they may be designed for multiple spans and multiple stories, as shown in Figure 3-21.

Steel Arches

In structural steel it is possible to design a two-hinged arch which, geometrically, would be a segment of a circle. A form such as this has great spanning ability and requires minimum amounts of material because of the efficient shape. Circular two-hinged arches can efficiently span from about 80 to 200 ft. Like the gabled or flat frame, it is also an indeterminate structure whose analysis is somewhat complex. As mentioned previously, however, the accessibility of computers today, even in small design offices, has eliminated the tedium experienced in hand computations. Arches may be made from plate stock welded together, but a more common form, perhaps, is the trussed arch, as shown in Figure 3-22.

It is possible to develop a great number of span, rise, and load combinations. When dealing with a circular arch, the architectural designer must be conscious of the internal clearances that may be required, as suggested in Figure 3-23, and model the arch based on these requirements. However, it may not be feasible to anchor a circular arch at the ground level because of enclosed space that may not be usable. Consequently, an arch may be supported by vertical supports, as shown in Figure 3-24. It must be emphasized, however, that the circular two-hinged arch will also have horizontal thrust forces at its base. Therefore, the walls or columns supporting such arches must be designed to resist overturning. The simplest way to avoid this problem is to use a tie-rod to connect the bases of the arch, as shown in Figure 3-25. Obviously, a tie-rod would be visually exposed and may not be desirable from either an aesthetic or a functional viewpoint.

Solid-web radial arch

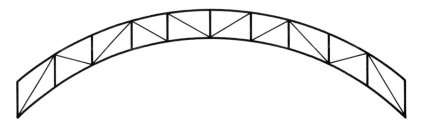

Trussed arch

Figure 3–22   Radial arches.

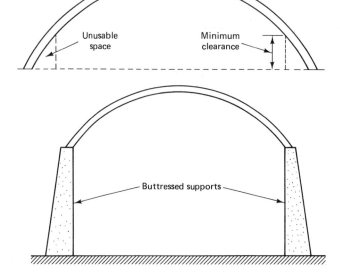

Unusable space

Minimum clearance

Figure 3–23   Radial arch and required clearances.

Buttressed supports

Figure 3–24   Radial arch on vertical supports.

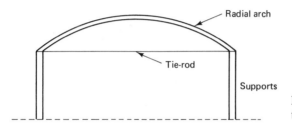

Figure 3–25 Radial arch with tie-rods.

Space Frames

In addition to the longspan structural systems discussed, there are a variety of other systems, although they are not as common. These include domes, steel-framed folded plates, cable-supported steel frames, lamella roofs, and space trusses (see Figure 3–26). Generally, a space frame is any three-dimensional structural frame where the parts are designed so that the entire frame acts as an integral structural unit. There are virtually an infinite variety of possibilities for the arrangement of parts for the development of a space frame. We limit our discussion here to the horizontal spanning space frame, which is a system of intersecting members that delivers load to supports in more than one direction. In particular, we concentrate on the space truss system.

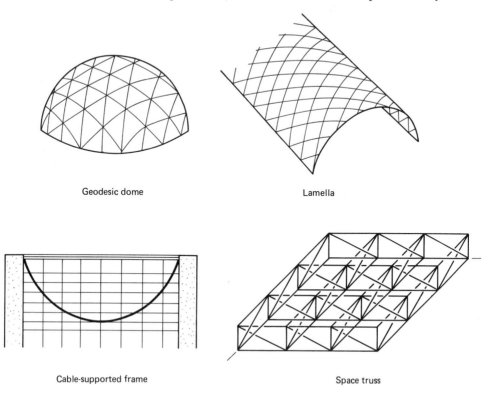

Figure 3–26

Space trusses can span great distances with minimum amounts of material, although the fabrication procedures are rather labor intensive. Where space trusses are used, it is not uncommon to see clear spans of well over 100 ft to, perhaps, 300 ft. In general, for a two-way space truss, the necessary minimum depth for efficiency may be estimated as about $\frac{1}{20}$ to $\frac{1}{24}$ of the span. The space truss may be supported by continuous supports, but more commonly they are supported by columns, which in themselves may be built of members arranged in a triangulated pattern, as shown in Figure 3–27. It is desirable, in a space truss structure, to have cantilevers that will introduce negative moments and, thereby, assist in minimizing deflection within the span. Ideally, the portion of the space truss that cantilevers should be about 25 to 30% of the clear span. Furthermore, since the space truss spans in two directions, the ideal situation would be to have cantilevers on all four sides of the structure. The number of supports required to support even very long spans may be as few as four, symmetrically placed at the corners. Ideally, space trusses should have equal clear spans in both directions.

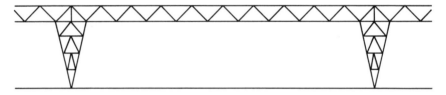

Figure 3–27   Space truss on triangulated supports.

There are a variety of ways to configure a space truss. The simplest space truss system may consist of intersecting flat Pratt trusses, as shown in Figure 3–28. A more efficient arrangement, however, may be the three-dimensional Warren truss. In a system such as this, the bottom chords of the truss are located midway between two top chords (see Figure 3–29). In other words, the bottom chord, which is the tension

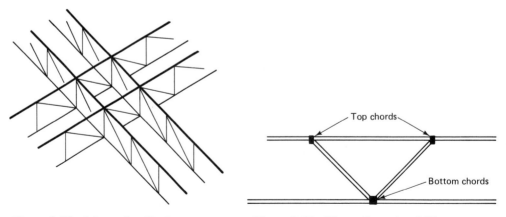

Figure 3–28   Intersecting Pratt   **Figure 3–29**   Three-dimensional Warren truss.

chord, would share two compression top chords. The web members connecting the top and bottom chords would then be arranged in a pyramidal fashion. For efficiency, the ideal situation is for the web members to be sloped at 45°. This would necessitate a space truss depth equal to one-half of the chord spacing. The chord spacing, ideally, should be the same in both directions. This spacing would be determined by the overall dimension of the building and the roofing system to be used. The roof system would be designed so that equal loads will be delivered to the intersections (sometimes called the *nodal* points). There are a variety of ways to do this, such as a two-way slab or a small space truss filling in the panels. Also, it is possible to use a space frame spanning between the nodal points, which may be pyramidal in shape, arched, and so on, as shown in Figure 3–30.

Where an absolutely flat roof is desired, the simplest and, perhaps, most efficient way to achieve equal load distribution to the chords is to use purlins and arrange them in a "checkerboard" fashion as shown in Figure 3–31. While the end

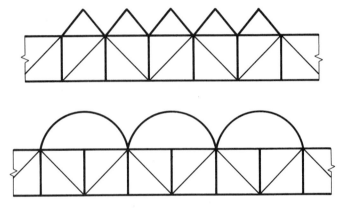

Figure 3–30

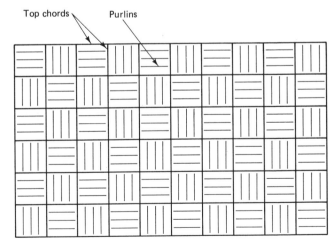

Figure 3–31 Checkerboard loading on space trusses.

trusses will not receive equal load distribution on the top chord, the load involved is small compared to the interior trusses and will not greatly affect the efficiency of the system.

Another longspan structural system worthy of note is the *lamella* system. Fundamentally, this is a system of intersecting skewed arches made of short members called lamellas (see Figure 3-32). The lamella system may also be used for a dome structure. The Louisiana Superdome in New Orleans, shown in Chapter 1 (Plate 1), is a lamella dome. The Superdome has a diameter of 680 ft. Where longspan domes are necessary, the lamella is an efficient structure because there is minimal bending stress involved, and the axial stresses are evenly distributed.

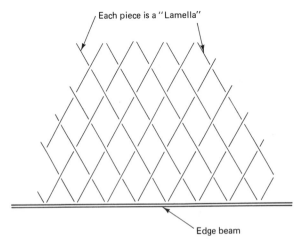

Each piece is a "Lamella"

Edge beam

Figure 3-32 Plan view—lamella roof.

STAGGERED TRUSS SYSTEM

An interesting possibility for a structural system where a high-rise building and long clearspan requirements are combined is the staggered truss system. Essentially, the system consists of trusses that are a full story in height and span between exterior columns. The trusses are staggered, vertically, as shown in Figure 3-33. Therefore, each truss provides support for two floors. Thereby allowing large column-free spaces. The only vertical obstructions would be the trusses themselves and these can be incorporated as part of the partitioning system. For lateral stability the floor system must be designed as a diaphragm in order to transfer lateral loads to the trusses. The truss diagonals deliver the lateral forces to the columns, where they are transferred as direct loads. Therefore, there will be no bending moments introduced to the columns, in the spanning direction of the truss. Lateral forces in the longitudinal direction, however, must still be accounted for, as they will produce bending in the column. This suggests that the columns can be oriented so that the strong axis will help to resist lateral forces in the longitudinal direction.

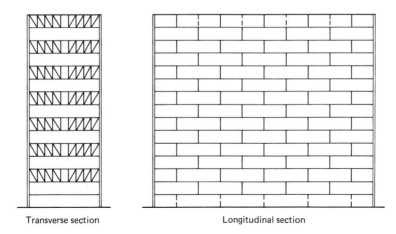

Transverse section   Longitudinal section

Figure 3-33 Staggered truss.

CONCLUSION

The word "systems" has been used many times throughout this book, especially in this chapter. It should be emphasized that, in building design, the number of systems that are possible should be considered as infinite. The possibilities are limited only by the creativity of the designer. For our purposes (building structures) a system may be defined as an arrangement of parts that interact and serve a common purpose. That purpose, of course, is to resist the forces to which the building will be subjected and to provide safety and comfort for the users.

The development of a system for a particular building often involves much deeper thought than may have been implied throughout this chapter. The designer must think in terms of the synthesis of all the parts that interact (and this includes more than the structural components): economy, constructability, and so on. Indeed, the few "systems" discussed previously, or elements of those systems, can be combined, thereby forming totally different systems.

SUPPLEMENTARY REFERENCES

COHEN, MICHAEL P. "Design Solutions Utilizing the Staggered-Steel Truss System," *AISC Engineering Journal,* Vol. 23, No. 3, Third Quarter 1986.

COLEMAN, ROBERT A. *Structural Systems Design.* Englewood Cliffs, N.J.: Prentice-Hall, Inc., 1983.

LIN, T. Y., and STOTESBURY, S. D. *Structural Concepts and Systems for Architects and Engineers.* New York: John Wiley & Sons, Inc., 1981.

SCHUELLER, WOLFGANG. *High-Rise Building Structures.* New York: John Wiley & Sons, Inc., 1977.

SCHUELLER, WOLFGANG. *Horizontal-Span Building Structures*. New York: John Wiley & Sons, Inc., 1983.

Space Forms in Steel, selected articles reprinted from *AISC Engineering Journal.* Chicago: American Institute of Steel Construction, 1965.

Staggered Truss Framing Systems for High-Rise Buildings, USS Technical Report ADUSS 27-5227-02. Pittsburgh, PA.: United States Steel, December 1972.

Steel Gables and Arches, AISC Publication T107. Chicago: American Institute of Steel Construction, 1963.

4

BEAMS—ALLOWABLE STRESS DESIGN

Now that we have completed the development of the necessary background information, we are ready to look at the process of designing members made of structural steel. We shall look at this process in both this chapter and the one following. In this chapter we are concerned only with beams: simple beams, cantilevers, and continuous beams. In the following chapter we will be concerned with the design of tension members and compression members. We follow the basic procedures as dictated by commonly accepted practices for the design of beams. This includes recommendations of the AISC (American Institute of Steel Construction).

In the beam design process there are three factors of importance for determining the size of the necessary structural steel beam for a given set of conditions. In order of priority, they are:

1. Design based on stress due to bending
2. Design based on deflection
3. Design based on shear

We will elaborate on these because in some cases the order of priority may necessarily be rearranged. In addition to these three major criteria, there are secondary issues in the design process that must also be addressed, and we will study these carefully.

Considering the three major points that must be evaluated in the design process, and to elaborate on the order of priority, which may be influenced by a variety of situations, let us consider point 1, design based on bending stress. It should be recalled from fundamentals that the use of the bending stress equation will be necessary for this, which is

$$F = \frac{Mc}{I}$$

where F = bending unit stress, psi or ksi
     M = bending moment on the section (we would normally be concerned with the maximum bending moment)
     c = distance from the neutral axis of the cross section to the level within the cross section being investigated (we would normally be concerned with investigating the stress at the outermost fibers)
     I = moment of inertia of the cross section

Depending on the background of the reader, the notations used here may be somewhat different from those originally learned, but the meaning of the bending stress equation, of course, has not changed.

The second factor, design based on deflection, is also one of great importance in the design process. Often, depending on the function of a support system, we would be greatly concerned with the amount of deflection (Δ) from the horizontal that would be produced by the loading, as shown in Figure 4–1. This concern could, in some instances, be more critical than bending stresses. It is not uncommon for the limiting amounts of deflection that would be tolerable to govern the final design of a beam. For example, if we were using a steel beam system to support a floor carrying sensitive machinery or equipment, such as mainframe computers, we would not want excessive vibrations or a "bouncy" floor system. This means that we would want a very stiff floor system. The amount of deflection (which is a measure of stiffness) that could be tolerated would therefore be quite small. Under such circumstances the design of the beam may be based on this very limited tolerance rather than on the bending stress. Consequently, bending stress and deflection, while listed in the order of priority that they were, may, under certain conditions, be reversed.

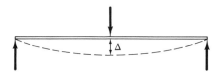

Figure 4–1 Deflection (Δ).

The third factor, shear, may or may not be an influencing factor in the design of a structural steel beam. As a general rule, in an architectural structure, it is of no great concern. The allowable shearing unit stresses for structural steel, as we will discuss in detail later, are extraordinarily high. Consequently, in the design of a steel beam, the usual procedure would be to design for allowable bending stress. If deflection limitations are an important issue, we would then investigate the member selected based on bending stress to see if it has the appropriate stiffness. If these concerns are satisfied, then, under "normal" circumstances, shear will be satisfied automatically because of the high allowable unit stress. In some cases, which generally do not occur in an architectural structure, shear may become the basis for the design of a beam. This would happen, for example, where we have very large loads on a very short span. Using the usual procedure of designing for bending stress based on the maximum bending moment, the short span would yield a relatively small value of moment and, therefore, a relatively small beam size. The shear at the end of the member, however, may be too high for the member sized. This same situation may also occur in a beam carrying very heavy loads close to the support. As a general rule such instances occur only in an architectural structure, where mechanical equipment is being supported. However, such conditions are not uncommon in industrial or heavy storage facilities, where very heavy loads may be supported in unusual ways. Figure 4–2 shows several situations where shear may be an issue of concern.

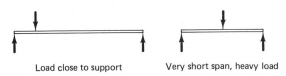

Load close to support   Very short span, heavy load

Figure 4–2 Conditions for critical shear.

DESIGN OF BEAMS

We will use the preceding discussion now to study the relatively simple procedure for designing a structural steel beam. As mentioned in Chapter 2, there are two possible methods for designing steel beams. They are the allowable stress design and plastic design methods. In this chapter we limit ourselves to the allowable stress design method, which is the most prevalent method. We start out with the design of simple beams, and then we look at cantilevers and continuous beams. We consider the three important criteria for design along with a discussion of AISC specification requirements and general recommendations.

Bending Stress

Before getting to the actual procedure for design, we must first make reference to certain requirements of the AISC Specification regarding the allowable stress that may be used in the design process. When considering bending there are essentially two possibilities for the allowable stress that may be used in the bending stress equation. They are

$$F_a = 0.60F_y \quad \text{or} \quad F_a = 0.66F_y$$

where F_a = maximum allowable unit stress in bending (tension or compression)

F_y = yield stress of the steel being used

The basis for the decision to use one or the other as the allowable stress is outlined very clearly in the AISC Specification. According to this, any section that qualifies as a "compact" section may be designed using the higher of the values given for the allowable bending stress; that is, $F_a = 0.66F_y$. For a section to qualify as a compact section there are a variety of criteria that must be satisfied.* Our primary concern will be with the design of wide-flange sections made of steel with $F_y = 36$ ksi (A 36 steel). For all practical purposes all of these sections qualify as compact sections under all of the applicable criteria, save one, which must be evaluated on a case-to-case basis.† This criterion deals with minimum requirements for the distance between lateral supports for the compression flange of the beam. It should be recalled from basics that thin elements in a state of compression have a tendency to buckle. The same is true in a steel wide-flange beam subjected to bending where one of the flanges is on the compression side. Unless there is a lateral support provided at certain intervals, there is a danger that the compression flange will buckle. The AISC Specification provides the basis for determining the maximum permissible interval for lateral support of the compression flange. If the compression flange is supported within this interval, the section qualifies as "compact." Fortunately, the values for these intervals have been calculated for each wide flange and we have these data available in the Appendix. Referring to Data Sheets A21 to A23 in the Appendix, there are two values given, shown as L_c and L_u. If the distance between lateral supports is less than that shown for L_c, the section qualifies as a compact section and the allowable bending stress may be taken as $0.66F_y$. If the actual distance between lateral supports is greater than L_c but less than the value shown for L_u, the section does not qualify as a compact section and the allowable bending stress is taken as $0.60F_y$. It is also possible to encounter a situation where the distance between lateral supports is greater than L_u. The allowable stress must then be further reduced. The AISC Specification gives equations for determining the usable allowable bending stress. However, a situation such as this is, it seems, highly unlikely in an architectural structure. In fact, it is recommended that the value of L_c not be exceeded, even if supplementary lateral bracing must be used.

For a simply supported beam in a building structure that is supporting a continuous floor slab, the distance between lateral supports is not an issue at all because the distance between lateral supports would be zero. This assumes minimum positive attachment (by whatever means) of the slab to the top flange of the beam. If, however, the beam is supporting, perhaps, the leg of a piece of equipment, with no floor slab resting on it, the distance between lateral supports for the compression flange would be the distance from the point load to the vertical support of the beam. As a general rule this sort of condition may be more likely to occur in an industrial facility.

When dealing with other than simply supported beams, such as a cantilevered beam, we must be very careful in our concern for the compression flange. In the main span the compression flange would, of course, be the top flange, for most of the span. However, when we consider the cantilever itself, it is the bottom flange that is the compression flange and, as a general rule, the bottom flange would not be supported laterally by a floor slab, which would be resting on the top flange, as

*These are detailed in Section 1.5.1.4 of the AISC Specification.

†For those who choose not to refer to the AISC Specification, let it suffice to say that there are five criteria applicable to wide-flange shapes, and four of these have to do with cross-sectional properties.

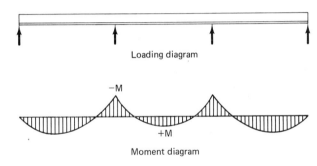

Loading diagram

−M

+M

Moment diagram

Figure 4-3 Bottom flange of cantilever in compression.

Figure 4-4 Continuous beam—bottom flange in compression at negative moment.

shown in Figure 4-3. The same situation is true for the distance from the support to the point of contraflexure in the main span. This condition also occurs in a continuous beam where we have negative and positive moments, as shown in Figure 4-4. In the region of negative moment the bottom flange is the compression flange. Therefore, in cantilevers and continuous beams a careful evaluation should be made for the distance between lateral supports of the compression flange. For a cantilever, the AISC Specification suggests that the unbraced length may be taken as the actual length provided that it is braced at the support. It is recommended, however, that the tip of the cantilever be laterally braced. In a building structure this is normally done as a matter of necessary detailing, such as a fascia piece spanning from cantilever to cantilever, as shown in Figure 4-5.

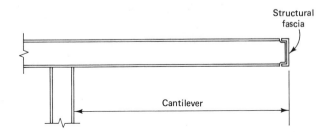

Structural fascia

Cantilever

Figure 4-5 Lateral bracing of cantilever.

Where, perhaps for architectural reasons, the full height of the tip of the cantilevers cannot be tied together, laterally, research shows (as strange as it seems) that the most effective bracing, *in the special case of a cantilever,* occurs when a single brace is placed on the top of the top flange. In addition to this, it is also recommended that an additional brace be placed on the compression side at about the midlength of the cantilever (see Figure 4-6).

A little more background information is necessary, and then we will learn to size a steel beam properly. For the purposes of this book we will, in every case, be concerned with selecting the most "economical" member. The most economical section, to do a given job, is defined here as the section with the least weight that satisfies

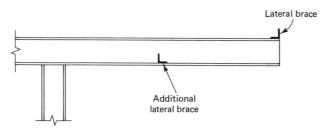

Figure 4-6 Additional bracing of cantilever.

the requirements for bending stress, deflection, and shear. To make this selection we will be using Data Sheets A5 to A14 in the Appendix. First, a few words are necessary regarding the meaning of the designations used for wide-flange sections. Steel sections are designated according to:

1. Shape of the section
2. Nominal depth (inches) of the section
3. Weight (lb/lin ft) of the section

For example, consider the following section designation, which may be found on the wide-flange shapes data in the Appendix (it is the first one given):

$$W36 \times 300$$

The various parts of this designation are defined as follows:

1. The symbol "W" indicates that this is a wide-flange shape. There are a variety of other shapes in steel (see Chapter 1), but we will be concerned primarily with the design of wide flanges.

2. The number "36" indicates the nominal depth of the section in inches. Nominal depth means that this section is made in the 36-in. rollers at the steel mill. The actual depth will usually be slightly different than the nominal depth.

3. The number "300" indicates the weight of the section in terms of pounds per linear foot. Since steel is priced by weight, this number is important in selecting the most economical section.

We are now ready to look at several examples, and in this context, we will further discuss and elaborate on many of the issues already presented. Reference will be made to the appropriate data sheets located in the Appendix.

Example 4-1

Select the most economical W shape to carry the given loading shown in Figure 4-7, using A36 steel and AISC Specifications. Consider that the compression flange is continuously laterally supported.

Solution   For this problem we will concern ourselves only with satisfying allowable bending stresses, since shear will not be an issue and no limitations are

Maximum bending moment:

$$M = \frac{WL}{8}$$

(see Data Sheet A2)

Simply supported beam   **Figure 4–7**

given for deflection. To do this we will use the bending stress equation (notations were defined earlier in the chapter):

$$F = \frac{Mc}{I}$$

The use of the equation in this form, however, presents a problem since c and I are both unknown in the design process. Therefore, we will alter this form of the equation by using a factor known as the *section modulus (S)*, where

$$S = \frac{I}{c} \text{ (units are in.}^3)$$

Therefore, the bending stress equation may be restated as follows:

$$F = \frac{M}{S} \quad \text{and} \quad S = \frac{M}{F}$$

In the design process we will select a member with a section modulus based on the full allowable stress (F_a). Based on the AISC Specification, $F_a = 0.66F_y = 24$ ksi.

$$M = \frac{WL}{8} = \frac{(48 \text{ kips})(24 \text{ ft})}{8} = 144 \text{ ft-kips}$$

$$S = \frac{M}{F_a} = \frac{(144 \text{ ft-kips})(12 \text{ in./ft})}{24 \text{ ksi}} = 72 \text{ in.}^3$$

Data Sheets A12 and A13 have the section moduli *(S)* tabulated for wide-flange beam sections and these data are arranged in a way to make the selection of the most economical (lightest weight) section a simple matter. In looking at the tables it will be noticed that the sections are arranged in distinct groups. At the top of each group is a section designated in bold type. To select the most economical section, simply find the section with the section modulus closest to that required, but not smaller. This will fall within one of the groups of sections. Then go to the top of the group to the section in bold type, and this will be the most economical section, even though it will normally have a section modulus greater than required. For example, in the problem being discussed here, we used $S = 72$ in.^{3}. In scanning the tables we find that the section closest to this is a W16 × 45, which has a section modulus of 72.7 in.[3]. If we go to the top of the group we will find that a W21 × 44 is the lightest section that will satisfy the requirements. Although it has a greater section modulus, this section weighs 1 lb/ft less. This design problem is now complete.*

In Example 4–1 we selected the most economical section based on least weight. It should be noted that this section is 5 in. deeper than the W16 × 45, which has a section modulus that would have satisfied the requirements. There may be occasions where we might wish to select the shallower section, albeit somewhat heavier. Although the cost of material would be greater, the shallower structural system may provide economy in other ways. For example, in a tall building the total height is, to a large degree, a function of the ceiling-to-floor thickness, which is usually controlled by clearances required for mechanical ductwork and depths of steel beams and girders. In a 20-story building, for example, if the structural depth were reduced, let us say by 6 in. per floor, the total building height would be reduced by 10 ft. Depending on the perimeter of the building, there would be a substantial reduction in the square footage of exterior finish required, which is an expensive item. Also, the height of columns, elevator shafts, and stairs will be reduced. Considering this, it may sometimes be difficult to determine the most economical system when looking at the overall picture. This question can only be answered by detailed analyses. Since, in our design examples, we have no specific building to deal with, the member with the least weight will be considered the most economical.

Example 4–2

In this example we wish to select the most economical section to carry the loading shown in Figure 4–8, using A36 steel and the AISC Specifications. The compression flange is laterally supported at the midspan and at the points of support. **Solution**   Assuming that the member qualifies as a compact section, then $F_a = 24$ ksi.

$$M = \frac{PL}{4} = \frac{(20 \text{ kips})(20 \text{ ft})}{4} = 100 \text{ ft-kips}$$

$$S = \frac{M}{F_a} = \frac{(100 \text{ ft-kips})(12 \text{ in./ft})}{24 \text{ ksi}} = 50 \text{ in.}^3$$

Referring to the Elastic Section Modulus Tables (Data Sheets A12 and A13), the most economical section is a W18 × 35. Because of the lack of lateral support of the compression flange for a length of 10 ft, we must determine if the use of $F_a = 24$ ksi was appropriate. The maximum distance between lateral supports to qualify as a compact section is given in Data Sheets A21 to A23. Referring to these data we see that a W18 × 35 has an $L_c = 6.3$ ft. Therefore, this section does not qualify as a compact section and, consequently, does

---

*In the Elastic Section Modulus Tables there is a column of information noted as F_y'. This is the theoretical yield stress at which the shape becomes noncompact, according to the AISC Specification. We will limit our designs to A36 steel and, consequently, we will not be concerned with these data.

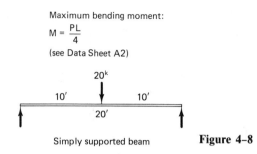

Maximum bending moment:

$M = \dfrac{PL}{4}$

(see Data Sheet A2)

Simply supported beam   **Figure 4–8**

not qualify for F_a = 24 ksi. The allowable stress must be reduced to $0.60F_y$ = 22 ksi. However, it will be seen in the Unbraced Length Tables that even at the reduced stress the section is not satisfactory since L_u = 6.7 ft. The only sensible solution to this dilemma, and one that is recommended when such a situation is encountered, is to provide supplementary lateral bracing to reduce the distance between lateral supports. In this case it would be sensible to provide such bracing at the quarter points of the span. This would reduce the distance between lateral supports to 5 ft and the section would qualify as a compact section. Considering that this approach is used, this design problem is complete.

Although it is difficult to imagine a situation in an architectural structure where compression flanges are laterally unsupported for long distances and where supplementary lateral bracing could not be used, the AISC Specification gives procedures for this sort of condition.* Laterally unsupported compression flanges are encountered, as mentioned earlier, in continuous beams and cantilevers in the negative moment region, where the bottom flange is the compression flange.

Example 4–3

Select the most economical wide flange beam to carry the loading shown in Figure 4–9, using A36 steel and the AISC Specification. Since this beam is a single cantilever, check the laterally unsupported distance for the compression flange. Assume that the floor system is providing lateral support for the top flange and that the support acts as a lateral brace for the compression flange.
Solution

$$S = \frac{M}{F_a} = \frac{(63.3 \text{ ft-kips})(12 \text{ in./ft})}{24 \text{ ksi}} = 31.6 \text{ in.}^3$$

Referring to the Elastic Section Modulus Tables (Data Sheets A12 and A13), the most economical section is a W12 × 26. The distance from the support to the point of contraflexure is, in this case, small and of no concern.† However,

*See Section 1.5.1.4.5 of the AISC Specification. Also, the AISC *Manual of Steel Construction* contains charts for the selection of beams with laterally unsupported compression flanges.

†For a refresher on locating points of contraflexure, see I. Engel, *Structural Principles* (Englewood Cliffs, N.J.: Prentice-Hall, Inc., 1984), Chapter 8.

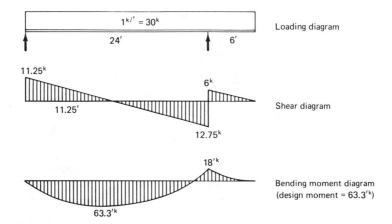

Figure 4-9　Single cantilever.

for the cantilever the compression flange is laterally unsupported for a distance of 6 ft. We assumed that the allowable stress was 24 ksi, which is true for a compact section. We must check to see if this qualifies as a compact section. Referring to the Unbraced Length Tables (Data Sheets A21 to A23), we see that the value of L_c for a W12 × 26 is 6.9 ft, which means that it qualifies as a compact section, and the design is complete.

It should be noted that in Example 4-3 the bending moment at the support is considerably lower than the design moment, which is the positive moment. The level of stress, therefore, in the compression flange at the cantilever is very low and, in any case, if the given value of L_c was less than that shown, the section would probably still be sufficient. If the given values of L_c or L_u were much smaller than those given, the AISC Specifications should be followed to determine the allowable stress. Supplementary lateral bracing may also be advisable in such a situation.

Example 4-4

Select a constant-section wide-flange beam for the continuous member shown in Figure 4-10, using A36 steel. Assume that the floor system provides continuous lateral support for the top flange and that the bottom flange is laterally supported only at the supports.

Solution　The critical moment for design is the negative moment of 57.6 ft-kips (which was determined using the three-moment theorem) and the unbraced length of the compression flange is 6.6 ft. Assuming an allowable stress of 24 ksi,

$$S = \frac{M}{F_a} = \frac{(57.6 \text{ ft-kips})(12 \text{ in./ft})}{24 \text{ ksi}} = 28.8 \text{ in.}^3$$

From the Elastic Section Modulus Tables (Data Sheets A12 and A13) we find that a W14 × 22 has a section modulus of 29.0 in.³. This is very close to the requirement and we have not included any consideration for the weight of the

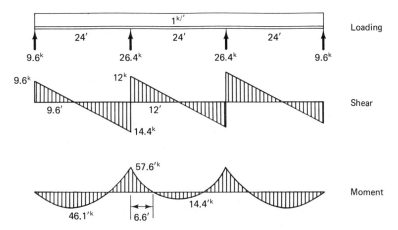

Figure 4-10   Continuous beam.

beam itself.* In addition, a check of the Unbraced Length Tables shows that a W14 × 22 has an L_c value of 5.3 ft. Consequently, the use of 24 ksi as the allowable stress is erroneous. In scanning the tables, it seems that the most sensible solution is to select a W12 × 26. The L_c value is 6.9 ft and the increased section modulus will take care of the beam weight. This design problem is now complete.

In the preceding examples we have concerned ourselves only with bending stress and directly related issues. In the following section we concentrate on the issue of deflection.

Deflection

The distance a beam bends from its original horizontal position, due to superimposed loading, is called *deflection*. The amount of deflection that can be tolerated in a particular situation is largely dictated by the intended function of the member and the judgment of the designer. Consideration of deflection, as mentioned earlier in this chapter, can be the governing criterion in the design of a beam. In many cases the tolerable deflection may be small enough so that the selection of a beam is governed by deflection rather than by the allowable bending stress.

The problems that can be caused by excessive deflection are multiple. They include possible damage to finishes due to cracking, water penetration when the exterior finish of a building is affected, glass breakage (or bowing if the pane of glass is large), and on flat roofs, excessive puddling (or ponding) of water, which causes increased load, thereby compounding the deflection problem. In addition to these

*Normally, the selection of the most economical section will provide enough extra section modulus to account for the weight of the beam. If the required and actual section moduli are very close, the beam weight should be considered.

problems there is the potential psychological problem of a "bouncy" floor system, which can cause some discomfort or, at best, some annoyance to the users of the building.

An appropriate question would be: How much deflection is tolerable in a beam? When are we concerned with allowable stresses we have strict guidelines provided by the AISC Specification. These guidelines are based on many years of testing and experience. However, the amount of deflection that can be tolerated in a steel beam is largely a matter of the designer's judgment regarding the necessary stiffness of a beam if it is to serve its intended function properly. There are a variety of recommendations, which have become traditional through the years, that can help guide the judgment of the designer. The most notable of these and one most frequently used is: *Deflection should not exceed 1/360 of the span (in inches), due to the live load.* This guideline was developed many years ago when plaster ceilings were common. It was determined that deflections due to live loading in excess of this would cause the plaster to crack. Although plaster ceilings are no longer common, this value for limiting deflection is still very much in general use and provides for a relatively stiff floor system. Sometimes this same criterion may be used based on the full loading, that is, dead load plus live load. This provides an even stricter limitation on deflection and therefore a stiffer floor system.

There is no reason, however, why the designer of a structure could not decide to tolerate more or less limiting deflections, depending on what is being supported. For example, limiting deflections may be taken from as large as 1/180 of the span (in inches) to as small, perhaps, as 1/1500 or 1/2000 of the span in inches. The larger value of limiting deflection (1/180) may be used for the roof system of a utilitarian sort of structure. However, it is not recommended that this be used on an absolutely flat roof system, even if it is thought that proper drainage is being provided. The smaller of the values (1/2000) may be necessary where the floor system supports very sensitive equipment that may malfunction due to the transfer of vibrations from the floor, which may be due to movement of people, or impact. Limiting deflection values between those given may be used depending on the function of the support system. Essentially, the key word is *judgment*.

In addition to the preceding guidelines, which generally may be applicable to any structure material, the AISC Specification also provides us with some guidelines. They are as follows:*

The most satisfactory solution must rest upon the sound judgment of qualified engineers. As a guide, the following rules are suggested:

1. The depth of fully stressed beams and girders in floors should, if practicable, be not less than $(F_y/800)$ times the span. If members of less depth are used, the unit stress in bending should be decreased in the same ratio as the depth is decreased from that recommended above.

*"Commentary on the AISC Specification (11/1/78)," *Manual of Steel Construction,* 8th ed. (Chicago: American Institute of Steel Construction, 1980), p. 5-139.

2. The depth of fully stressed roof purlins should, if practicable, be not less than $(F_y/1000)$ times the span, except in the case of flat roofs.

Ponding on roofs is an issue related to deflection and must be given serious consideration. Ponding causes increased loads due to the retention of water on flat-roof buildings. Ponding will occur due to low spots in the roof caused by initial deflection due to the roof loads, and the ponding will therefore increase the load, causing still further deflection. As the reader can imagine, this problem will continue to compound itself. If the framing system lacks sufficient stiffness, the accumulated weight can actually result in collapse. This situation may be particularly severe in areas where ponded water may turn to ice, thereby blocking drains and causing further buildup.

The Commentary of the AISC Specification provides formulas for estimating the weight and distribution of ponded water. Rather than resort, however, to the use of the fairly complex formulas, it is strongly recommended that roofs which are *absolutely* flat be avoided by the architectural designer. The roof should have some slope to one edge of the building to which water can run and be properly drained. Visually, this should present no problem. For example, imagine a building, say 30, 40 or 50 ft wide or more, and an architectural decision to have a flat roof. Even over minimum widths, the roof can slope, from one edge to the other, several inches and still, perceptibly, be a flat roof. This presents no problems in terms of the aesthetic decision by the architectural designer, and it presents few, if any, problems in the detailing of the building. In any case, it is strongly recommended that, by whatever means, water ponding be avoided.

We now proceed to several example problems where we will investigate deflection in steel beams. For this purpose, reference will be made to Data Sheet A2, which provides us with general expressions for determining deflections for a variety of loads. Again, it is assumed, as it is throughout this book, that the student has a background in the basics of structural mechanics.

Example 4–5

Select a wide-flange member to carry the given loading, as shown in Figure 4–11, using A36 steel. The maximum tolerable deflection is limited to $1/300$ of the span (in inches), under full loading.

Solution   We will assume, for the purpose of this problem, that the compression flange is continuously laterally supported.

$$M = \frac{WL}{8} + \frac{PL}{3} = \frac{(48 \text{ kips})(24 \text{ ft})}{8} = \frac{(4 \text{ kips})(24 \text{ ft})}{3} = 176 \text{ ft-kips}$$

$$S = \frac{M}{F_a} = \frac{(176 \text{ ft-rips})(24 \text{ in./ft})}{24 \text{ ksi}} = 88 \text{ in.}^3$$

Referring to the Elastic Section Modulus Tables (Data Sheets A12 and A13), a W18 × 50 could be used, assuming that the loading includes some allowances for the beam weight. A W21 × 50, which has the same weight, could also be used. If the 3 in. of additional depth presents no problem, this would be the

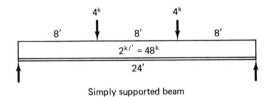

Maximum bending moment:

$$M = \frac{WL}{8} + \frac{PL}{3}$$

(see Data Sheet A2)

Simply supported beam   **Figure 4-11**

preferable choice. Referring to the Wide-Flange Shapes Tables (Data Sheets A5 to A11), we find that a W21 × 50 has a moment of inertia of 984 in.[4]. To check deflection we can use the general expressions given on Data Sheet A2 and add the results. Therefore,

$$\Delta = \frac{5WL^3}{384EI} + \frac{23PL^3}{648EI} \quad (E = 29{,}000 \text{ ksi})$$

$$\Delta = \frac{5(48)(24 \times 12)^3}{384(29{,}000)(984)} + \frac{23(4)(24 \times 12)^3}{648(29{,}000)(984)} = 0.64 \text{ in.}$$

The allowable deflection is

$$\Delta \leq \frac{L}{300} = \frac{(24 \text{ ft})(12 \text{ in./ft})}{300} = 0.96 \text{ in.}$$

Therefore, the member selected based on bending stress satisfies the deflection criterion, and the solution is complete.

Had the member selected for bending not been satisfactory for deflection, the deflection expression would be equated to the tolerable deflection value and a required moment of inertia would be determined. An appropriate member could then be selected from the Moment of Inertia Selection Table (Data Sheet A14).

Example 4-6

Select a constant section wide flange to carry the loading shown in Figure 4–12 using A36 steel. Also, determine the deflection at the midspan and at the tip of the cantilever on the right side. Assume that the top flange is continuously laterally supported. The modulus of elasticity *(E)* = 29,000 ksi.

Solution   Based on the moment diagram, the design moment is 48 ft-kips. Therefore,

$$S = \frac{M}{F_a} = \frac{(48 \text{ ft-kips})(12 \text{ in./ft})}{24 \text{ ksi}} = 24 \text{ in.}^3 \quad \text{use W14} \times 22$$

The value of Lc is shown as 5.3 ft for this member. The cantilever on the right side has an unsupported compression flange for 6 ft. However, the moment in the cantilever (and consequently, the level of stress in the flange) is only half

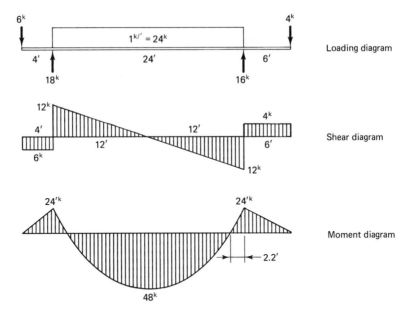

Figure 4-12   Double cantilever.

of the moment for which the member was designed. Since the values of Lc are based on a stress of 24 ksi, the 6-ft length of the unsupported compression flange on the cantilever is all right, considering the very low level of stress. In any case, it is recommended that, at least, the tip of the cantilever be braced.

To determine the deflection at the midspan we will use the principle of superposition. To begin, consider the main span as a simply supported beam, as shown in Figure 4-13a, and compute the deflection for this condition (W14 × 22; $I = 199$ in.[4]). Then superimpose the effect of the negative moments at the supports, as shown in Figure 4-13b.* The net deflection at the midspan is 0.77 in.

To determine the net deflection at the tip of the cantilever on the right side, we must first determine the net slope ($EI\theta$) of a tangent to the elastic curve at the right-hand support. The general expression for $EI\theta$ is given on Data Sheet A2. The computations are shown as part of Figure 4-13. It can be seen from this that the net value of $EI\theta$ at the right support is 288 kip-ft^2, as shown in Figure 4-14a. The deflection at the tip of the cantilever can now be computed by visualizing a fixed the ($EI\theta = 0$) as shown in Figure 4-14b and then adjusting the deflection by rotating the support through the value of $EI\theta$ as shown in Figure 4-14c. The deflection for a cantilever with a fixed end is given on Data Sheet A2. Therefore, the net deflection (Δ) at the tip of the cantilever is

*The expression for slope ($EI\theta$) and deflection shown in Figure 4-13b are based on a member subjected to equal end moments. The reader can derive these using the principles of slope and deflection.

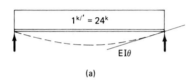

$$\Delta = \frac{5WL^3}{384EI} = \frac{5(24)(24 \times 12)^3}{384(29,000)(199)} = 1.29 \text{ in.}$$

$$EI\theta = \frac{WL^2}{24} = \frac{(24)(24)^2}{24} = 576 \text{ kip-ft}^2$$

(a)

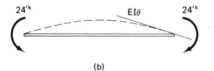

$$\Delta = \frac{ML^2}{8EI} = \frac{(24 \times 12)(24 \times 12)^2}{8(29,000)(199)} = -0.52 \text{ in.}$$

$$EI\theta = \frac{ML}{2} = \frac{(24)(24)}{2} = -288 \text{ kip-ft}^2$$

(b)

Figure 4–13

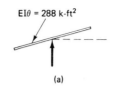

$EI\theta = 288 \text{ k-ft}^2$

(a)

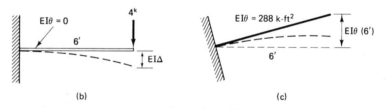

$EI\theta = 0$ 4^k $EI\theta = 288 \text{ k-ft}^2$

6′ $EI\theta (6')$

$EI\Delta$ 6′

(b)   (c)

Figure 4–14

$$EI\Delta = \frac{PL^3}{3} - EI\theta(6) = \frac{(4 \text{ kips})(6 \text{ ft})^3}{3} - (288)(6) = -1440 \text{ kip-ft}^3$$

The negative sign indicates a net upward deflection.

$$\Delta = \frac{(1440 \text{ kip-ft}^3)(12 \text{ in./ft})^3}{(29,000 \text{ ksi})(199 \text{ in.}^4)} = 0.43 \text{ in. (above the horizontal)}$$

Example 4–7

Design a constant-section wide-flange continuous beam to carry the loading shown in Figure 4–15. Determine the deflection in each span. Use A36 steel and the AISC Specifications.

Solution   The maximum bending moment is 104 ft-kips and, therefore,

$$S = \frac{M}{F_a} = \frac{(104 \text{ ft-kips})(12 \text{ in./ft})}{24 \text{ ksi}} = 52 \text{ in.}^3$$

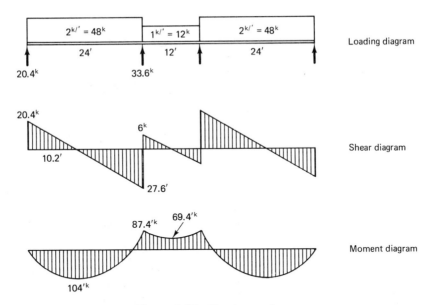

Figure 4-15 Continuous beam.

Referring to the Elastic Modulus Tables (Data Sheets A12 and A13), it is determined that the most economical section is a W18 × 35. The value of Lc for this section is given as 6.3 ft. It should be noted, from the moment diagram, that the middle span is totally in a state of negative bending, meaning that the bottom flange is the compression flange. It is recommended that this be remedied with the placement of, at least, one line of lateral bracing.

The deflection in the middle span is easy to determine accurately by the process of superposition. This is shown in Figure 4-16, along with the com-

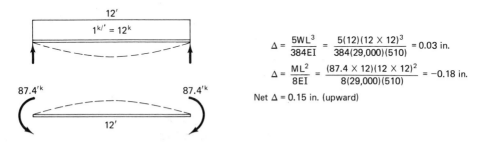

$$\Delta = \frac{5WL^3}{384EI} = \frac{5(12)(12 \times 12)^3}{384(29,000)(510)} = 0.03 \text{ in.}$$

$$\Delta = \frac{ML^2}{8EI} = \frac{(87.4 \times 12)(12 \times 12)^2}{8(29,000)(510)} = -0.18 \text{ in.}$$

Net $\Delta = 0.15$ in. (upward)

Figure 4-16 Deflection in middle span.

putations. The midspan deflection in the end span may also be determined by the process of superposition, even though the deflected shape is unsymmetrical.* This procedure is shown in Figure 4-17.

*The *midspan* deflection ($EI\Delta$) produced by a moment at one end is $ML^2/16$.

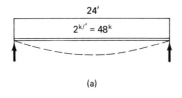

(a)

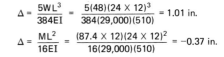

$$\Delta = \frac{5WL^3}{384EI} = \frac{5(48)(24 \times 12)^3}{384(29,000)(510)} = 1.01 \text{ in.}$$

$$\Delta = \frac{ML^2}{16EI} = \frac{(87.4 \times 12)(24 \times 12)^2}{16(29,000)(510)} = -0.37 \text{ in.}$$

Net $\Delta = 0.64$ in. (downward)

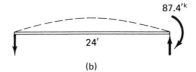

(b)

Figure 4-17   Deflection in end span.

It should be noted that the deflection in the short span is upward. This will occur frequently in short spans that are flanked by long spans, especially when the long spans are heavily loaded. Although we have not discussed limits for upward deflection, such conditions can produce problems if the amount of deflection is large. In any case, the designer should be aware of such conditions and make appropriate judgments about the potential consequences.

Shear

As mentioned in the introduction to this chapter, shearing unit stress is seldom an issue of great concern when building with steel due to the relatively high allowable stress. The AISC Specification recommends that on the cross-sectional area resisting shear, the allowable shearing unit stress be taken as

$$F_v = 0.40F_y$$

where F_v = allowable shearing unit stress
F_y = yield stress of the material

For the most common grade of steel used in an architectural structure (F_y = 36 ksi), this means that the allowable shearing unit stress is

$$F_v = 0.40(36 \text{ ksi}) = 14.4 \text{ ksi}$$

To determine the *actual* shearing unit stress in a given situation, it should be recalled, from basic structural mechanics, that the shearing unit stress equation is as follows:

$$F_v = \frac{VQ}{IB}$$

where V = total shear on the section being investigated
Q = first moment of area ($A\bar{x}$) of the cross section, above or below the plane within the cross section where the shearing unit stress is being investigated

    I = moment of inertia of the cross section

    B = width of the cross section at the plane being investigated

It should also be recalled that the shearing unit stress formula will give us a parabolic variation in the shearing unit stress throughout the cross section, with the stress being zero at the outermost fibers and maximum at the neutral axis. In every case our only concern will be with the maximum value.

    The application of the shearing unit stress formula for a wide-flange section, although not particularly difficult, would be somewhat of a nuisance. Fortunately, the AISC Specification recommends that, for rolled wide-flange sections, the effective area for resisting shear may be taken as the product of the depth of the section times the web thickness. What, in fact, this suggests is that the web alone will resist the shear, and that we may assume the shearing unit stress to be *averaged* over the area of the web, rather than a parabolic variation. In fact, this assumption gives results that are very close to the actual shearing unit stress that would be determined if we used the full cross section and the basic shearing unit stress equation.

    Although it has been suggested, on several occasions, that shear is normally of minimum importance, it seems appropriate at this time to present an example to, at least, show how one would check for shearing unit stress in accordance with AISC recommendations. We will do this following the normal design procedure of first selecting a member based on the allowable bending stress and then investigating the shearing unit stress for the member selected.

Example 4–8

    Select a wide-flange beam to carry the loading shown in Figure 4–18, using A36 steel and the AISC Specification. Because there is a heavy concentrated load close to one end, it is advisable to check the shearing stress.

    Solution   We will first select the most economical member based on bending stress and then investigate the shearing stress. As shown in the moment diagram, the maximum moment is 234 ft-kips. Therefore,

$$S = \frac{M}{F_a} = \frac{(234 \text{ ft-kips}) (12 \text{ in./ft})}{24 \text{ ksi}} = 117 \text{ in.}^3 \qquad \text{use W21} \times 62$$

Referring to Data Sheets A5 to A11, we can find the information necessary to check shearing stress. The depth of the section d is 20.99 in. and the web thickness (t_w) is 0.40 in. The allowable shearing stress (F_v) based on A36 steel is 14.4 ksi. The actual shearing stress (f_v) is simply the total shear divided by the area of the web.

$$f_v = \frac{42 \text{ kips}}{(20.99 \text{ in.})(0.40 \text{ in.})} = 5 \text{ ksi} < 14.4 \text{ ksi}$$

Even with a heavy load placed relatively close to the support, the selected member is very safe for shearing stress, and this problem is complete.

    In Example 4–8 it was assumed that the full area of the web was intact. However, if the flange of the beam is coped or if bolt holes exist in the web, this would

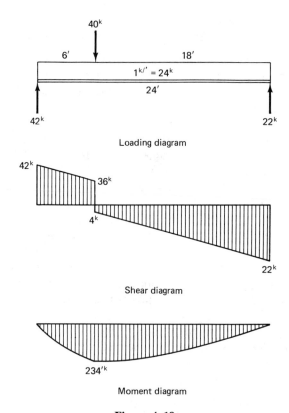

Figure 4-18

not be the case. For such conditions, where shear appears to be critical, reference should be made to the AISC Specification for recommendations and procedures.

LOADING ARRANGEMENTS

When analyzing cantilevers and continuous beams to determine the maximum moment for which the beam is to be designed, there is a further issue of which the student should be aware. This has to do with the fact that there are negative moments at the supports. In a cantilevered beam, the value of the negative moment, which is a function of the loading on the cantilever, will have an effect on the positive moment in the main span. Similarly, in a continuous beam, the loads on each span will have an effect on the maximum moments produced in the adjacent spans. In the examples presented thus far, the loads on cantilevered beams and continuous beams have been given as total loads. The total loads are, in fact, made up of dead loads and live loads. Only the dead loads are predictable in terms of permanence. Live loads, on the other hand, may or may not be present at any given time. Therefore, in a cantilevered beam, for example, if only the dead load was present on the cantilever and the full load was present on the main span, the maximum positive moment would be greater than that produced by full loading throughout (see Figure 4–19).

In a continuous beam, to analyze for maximum moments, we separate the dead and live loads and use a "checkerboard" loading pattern to determine the maximum moments. For example, in the three-span continuous beam shown in Figure 4–20, an analysis based on full loading throughout would yield certain values for negative and positive moments. If we now considered the fact that the live load may not be present on the middle span, the loading pattern would be shown in Figure 4–21, and this loading condition would produce higher positive moments in the end spans. If we reversed the loading pattern, as shown in Figure 4–22, the positive moment in the middle span would be greater than that found for full loading throughout. When analyzing for the critical positive moments in a continuous beam, the following rule should be used: *To determine the maximum positive moment in a span, load that span and alternate spans on each side with the total load. The spans in between are loaded with only the dead load.*

Because of the alternating pattern of full load and dead load, we often refer to this as checkerboard loading. Alternating load patterns on a continuous beam will

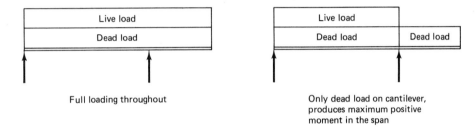

Full loading throughout

Only dead load on cantilever, produces maximum positive moment in the span

Figure 4–19

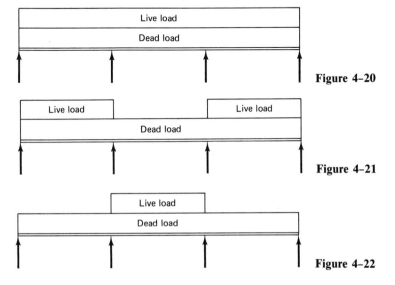

Figure 4–20

Figure 4–21

Figure 4–22

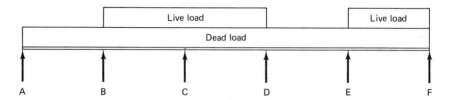

Figure 4-23   Loading for maximum negative moment at support C.

also have an effect on the negative moment at a particular support. To determine the maximum negative moment at a support, use the following rule: *Use full load on spans adjacent to the support and on alternate spans beyond. Load all other spans with dead load only (see Figure 4-23).*

An analysis considering the possible variations of full load and dead load only may actually reveal useful information regarding the deflection that may occur due to possible load variations. We elaborate on these ideas in the context of an example.

Example 4-9

Determine the most economical section required to carry the loading shown on the double cantilevered beam of Figure 4-24a. Consider critical arrangements of dead load and live load. Also determine the maximum possible deflec-

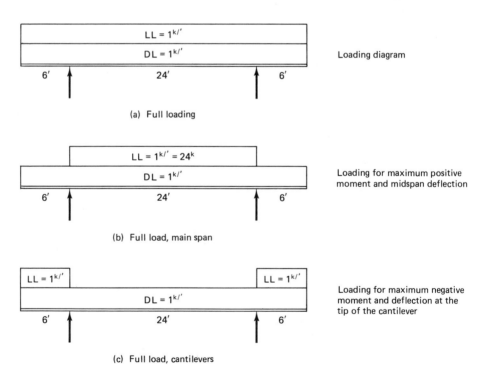

Figure 4-24   Loading arrangements on double cantilever.

tion at the midspan and at the tip of the cantilever. Use A36 steel and the AISC Specification.

Solution Based on the loading arrangements shown in Figure 4–24b and c the design moment is the positive moment of 126 ft-kips. Therefore,

$$S = \frac{M}{F_a} = \frac{(126 \text{ ft-kips})(12 \text{ in./ft})}{24 \text{ ksi}} = 63 \text{ in.}^3 \qquad \text{use W16} \times 40$$

The selected member has an L_c value of 7.4 ft; therefore, no supplementary bracing is required. We will now proceed to determine the deflection at the midspan and at the tip of the cantilever. The member selected (W16 × 40) has a moment of inertia of 518 in.[4]. The greatest deflection at the midspan will occur under the loading arrangement shown in Figure 4–24b. The deflection may be determined by superimposing the values of deflection for a simply supported beam and the effect of the negative moments at the supports. This is shown in Figure 4–25 along with the necessary computations. The computations shown represent the greatest possible deflection that can occur within the span, considering critical arrangements of dead and live loads.

In order to determine the greatest possible deflection at the tip of the cantilever, we must first determine the value of the slope of a tangent ($EI\theta$) to the elastic curve, at the support, based on the loading shown in Figure 4–24c. These conditions are shown in Figure 4–26. Therefore, using the same process as in Example 4–6, the maximum deflection at the tip of the cantilever may be determined by superimposing the values shown in Figure 4–27.

The deflection ($EI\Delta$) can now be computed by adding the results of Figure 4–27b and c. Therefore,

$$EI\Delta = \frac{WL^3}{8} - EI\theta(6 \text{ ft}) = \frac{(12 \text{ kips})(6 \text{ ft})^3}{8} - (144)(6) = -540 \text{ kip-ft}^3$$

The negative value indicates an upward deflection. The net value of deflection is

$$\Delta = \frac{(540 \text{ kip-ft}^3)(12 \text{ in./ft})^3}{(29,000 \text{ ksi})(518 \text{ in.}^4)} = 0.06 \text{ in. (above the horizontal)}$$

This problem is now complete.

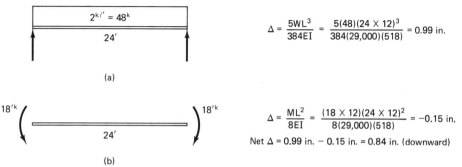

(a)

$$\Delta = \frac{5WL^3}{384EI} = \frac{5(48)(24 \times 12)^3}{384(29,000)(518)} = 0.99 \text{ in.}$$

(b)

$$\Delta = \frac{ML^2}{8EI} = \frac{(18 \times 12)(24 \times 12)^2}{8(29,000)(518)} = -0.15 \text{ in.}$$

Net $\Delta = 0.99$ in. $- 0.15$ in. $= 0.84$ in. (downward)

Figure 4–25 Maximum midspan deflection.

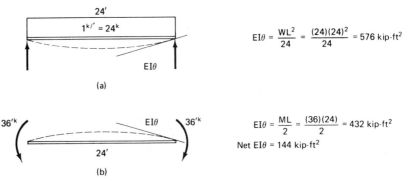

$$EI\theta = \frac{WL^2}{24} = \frac{(24)(24)^2}{24} = 576 \text{ kip-ft}^2$$

(a)

$$EI\theta = \frac{ML}{2} = \frac{(36)(24)}{2} = 432 \text{ kip-ft}^2$$

Net $EI\theta = 144$ kip-ft^2

(b)

Figure 4-26 Slope at the supports.

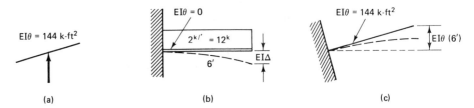

(a)   (b)   (c)

Figure 4-27 Deflection at tip of the cantilever.

COMPOSITE SECTIONS

When cast-in-place concrete floor slabs are supported on steel beams, the beams may be designed as "composite sections." In this type of construction, both materials are designed to act in unison by creating a positive bond between them. The concrete is therefore considered to have structural capability rather than simply being a dead weight on the steel beam. A typical cross section of a composite beam is shown in Figure 4–28. Composite action may be achieved between the concrete slab and the

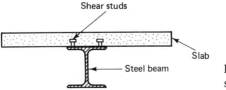

Figure 4-28 Typical composite section.

steel beam through the use of "shear studs." The shear studs, which are like steel "mushrooms" welded to the top flange of the beam, are required to resist the tendency of the slab to slide with respect to the beam when deformed under the influence of a load. This is an example of horizontal shear which must be resisted so that the beam and slab will act as an integral structural unit (see Figure 4–29). When dealing

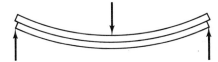

Figure 4-29 Beam deformed under load. Horizontal shear produces the tendency for separation along horizontal planes.

with composite construction we must be able to determine how much of the slab works effectively with the steel beam. For example, in the section shown in Figure 4-30, only a certain portion of the slab will act as an integral flange on the steel beam. Although it is difficult to determine the amount of the slab that is effective, in absolute terms, there are certain recommendations provided by the steel industry which will serve as guidelines. These recommendations are as follows (see Figure 4-31).*

Figure 4-30 Section—basis for determination of effective flange width.

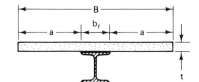

1. $B = \frac{1}{4}$ beam span

2. $a = \frac{1}{2}$ clear span between beams
   Total width (B) = 2a + b_f

3. a = 8 times the slab thickness (t)
   Total width (B) = 2a + b_f

Figure 4-31 Effective flange width.

When the slab extends on both sides of the steel beam:

1. The effective width of the concrete flange is to be taken as not more than one-fourth of the beam span.

2. The effective projection beyond the edge of the beam is to be not more than one-half the clear distance to the next beam.

3. The effective projection beyond the edge of the beam is to be not more than eight times the slab thickness. Normally, this criterion will govern the effective width of the slab.

When these recommendations have been evaluated, the least value shall be used as the effective width of the slab.

When the slab exists on only one side of the beam (as in a spandrel beam), the criteria are as follows:

1. The effective width of the concrete flange (the projection beyond the beam flange) is to be taken as not more than one-twelfth of the beam span.

*Where steel deck is used, the AISC Specification suggests that the total slab thickness, including ribs, be used when determining the effective width of the slab. See Section 1.11.5 of the AISC Specification for restrictions and requirements.

2. The projection of the slab beyond the beam flange is to be not more than six times the slab thickness.

3. The projection of the slab beyond the beam flange is to be not more than one-half the clear distance to the adjacent beam.

Again, the least of the values obtained from the evaluation of these criteria is to be used in the design of a composite section.

To analyze a composite section, the transformed section concept must be applied.* In the transformed section principle we transform the concrete slab to an equivalent area of steel. This is done by reducing the effective width of the slab by the modular ratio (n), where

$$n = \frac{\text{modulus of elasticity of steel}}{\text{modulus of elasticity of concrete}} = \frac{E_s}{E_c}$$

The modulus of elasticity of steel, regardless of grade, is taken as 29×10^3 ksi. The modulus of elasticity of the concrete varies depending on the quality of the concrete mix. Modular ratios (n) for the most common concrete mixes, based on the ultimate strength (f'_c) of the concrete are listed in Table 4-1.

TABLE 4-1 MODULAR RATIOS: VALUES OF E_c AND n FOR VARIOUS f'_c

f'_c (psi)	$n = E_s/E_c$	E_c (ksi)
3000	9	2900
3500	8.5	3450
4000	8	3650
4500	7.5	3900
5000	7	4100

Composite sections offer some advantages in terms of material savings. To demonstrate the advantages involved, we will proceed to an example. Before doing so, however, it is important to mention that there are a number of complexities involved in the analysis or design of a composite section which are not being presented here. The following example is being presented for the sole purpose of indicating, to the student, the inherent advantages of composite construction. For needs beyond this, it is recommended that reference be made to the AISC Specification.†

---

*For a refresher in the use and application of the transformed section principle, see I. Engel, *Structural Principles* (Englewood Cliffs, N.J.: Prentice-Hall, Inc., 1984).

†See AISC Specification in the *Manual of Steel Construction*, 8th ed., Section 1.11. The Specification provides strict guidelines for design when the construction is shored or unshored.

Example 4–10

For the framing plan shown in Figure 4–32, determine the most economical wide-flange section required for the beam indicated as B1, without benefit of composite action. Using the selected section, determine the uniform load-carrying capacity considering composite action between the beam and slab. Use A36 steel and concrete with ultimate strength (f'_c) of 3000 psi. The allowable stress for concrete (f_c) is $0.45 f'_c = 1350$ psi. The modular ratio is based on the concrete being transformed to an equivalent area of steel. Therefore, the reciprocal of the values shown in Table 4–1 is used, and $n = \frac{1}{9}$.

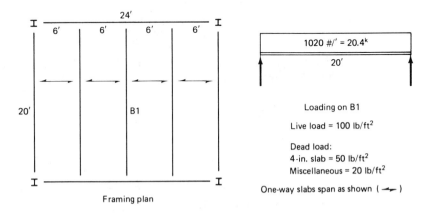

Figure 4–32   Framing plan analysis.

Solution   Referring to the loading on B1 shown in Figure 4–32, we have

$$M = \frac{WL}{8} = \frac{(20.4 \text{ kips})(20 \text{ ft})}{8} = 51 \text{ ft-kips}, \ S = 25.5 \text{ in.}^3, \text{ use W14} \times 22$$

This is the most economical section without composite action. We will now determine the uniform load-carrying capacity of this section with composite action. The necessary data for a W14 × 22 is $I = 199$ in.^{4}, $A = 6.49$ in.^{2}, $d = 13.74$ in., and $b_f = 5.0$ in. To determine the effective width of the slab that will act with the beam, we must test the three criteria given previously.

1. The effective width of the concrete slab is to be not more than one-fourth of the beam span: $(\frac{1}{4})(20 \text{ ft}) = 5 \text{ ft} = 60$ in.
2. The effective projection of the concrete slab beyond the edge of the beam is to be not more than one-half the clear distance to the next beam: $(\frac{1}{2})$ (6 ft)(2) = 6 ft = 72 in.
3. The effective projection of the concrete flange beyond the edge of the beam is to be not more than eight times the slab thickness: 8(4 in.)(2) + 5 in. = 69 in.

The first criterion governs and the composite section is shown in Figure 4–33, along with the transformed section. The concrete was transformed by the modular ratio ($n = \frac{1}{9}$) to an equivalent area of steel.

We must now determine the location of the neutral axis of the transformed section.

ΣAx (using the top of the figure as the reference):

$$(4 \text{ in.})(6.7 \text{ in.})(2 \text{ in.}) = 53.6 \text{ in.}^3$$
$$(6.49 \text{ in.}^2)(10.87 \text{ in.}) = 70.6 \text{ in.}^3$$
$$\Sigma Ax = 124.2 \text{ in.}^3$$

$$\bar{x} = \frac{\Sigma Ax}{\Sigma A} = \frac{124.2 \text{ in.}^3}{33.3 \text{ in.}^2} = 3.7 \text{ in.}$$

We will now determine the transformed moment of inertia (I_{tr}), ignoring any concrete below the neutral axis since this is the tension side of the section.

I_{tr}:

$$\frac{BD^3}{3} = \frac{(6.7 \text{ in.})(3.7 \text{ in.})^3}{3} = 113.1 \text{ in.}^4$$
$$199 \text{ in.}^4 + (6.49 \text{ in.})(7.1 \text{ in.})^2 = 526.2 \text{ in.}^4$$
$$I_{tr} = 639.3 \text{ in.}^4$$

In order to determine the maximum uniformly distributed load that may be carried, we will use the bending stress equation.

$$M = \frac{WL}{8} = \frac{F_a I_{tr}}{c}$$

Steel: $\dfrac{W(20 \text{ ft} \times 12 \text{ in./ft})}{8} = \dfrac{(24 \text{ ksi})(639.3 \text{ in.}^4)}{14 \text{ in.}}$ $W = 36.5$ kips

Concrete: $\dfrac{W(20 \text{ ft} \times 12 \text{ in./ft})}{8} = \dfrac{(1.35 \text{ ksi})(639.3 \text{ in.}^4)}{(3.7 \text{ in.})(\frac{1}{9})}$ $W = 70$ kips

Therefore, the maximum uniformly distributed load that can be carried by the W14 × 22 acting compositely with the slab is 36.5 kips. This is a substantially greater load than can be carried as a noncomposite section. This should serve to point out the primary advantages of composite construction, and the problem is complete.

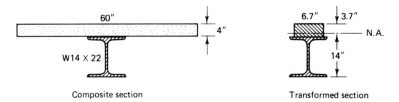

Composite section   Transformed section

Figure 4–33

In addition to the greater load-carrying capacity provided by composite action, as suggested by Example 4–10, there will also be greater stiffness since the moment of inertia used for deflection computations is the transformed moment of inertia. There will also be an advantage provided by shallower depths of members. All of these advantages can lead to significant advantages in economy.

OTHER SHAPES

In all of the preceding discussions and examples, we have concerned ourselves only with the selection of wide-flange sections. There are, indeed, other sections made of structural steel, as indicated in Chapter 1. It is not at all uncommon to use channel sections, angles, and "tee" sections in a structure. In steel construction, these shapes are normally not used as major supporting members. For example, channels are often used as stair stringers or structural fascias, as shown in Figure 4–34. Angles and tees are commonly used as lintels over openings where a masonry wall must be supported (see Figure 4–35).

One major concern when using unsymmetrical sections such as channels or angles is the problem of torsion. In wide-flange sections the usual process of design for bending is based on the idea that no twisting will take place, because we are dealing with a section having two axes of symmetry. In such sections all stresses are symmetrical about the intersection of the two axes of symmetry, which coincide with the centroid of the cross section. However, when using a cross section with only one axis of symmetry, this is not the case. In an unsymmetrical section, the center of rotation, commonly called the *shear center,* is the point through which the bending axis passes.

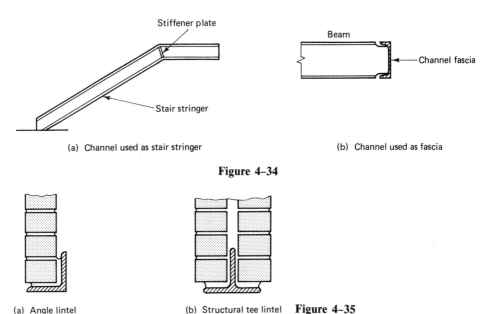

Stiffener plate

Beam

Channel fascia

Stair stringer

(a) Channel used as stair stringer

(b) Channel used as fascia

Figure 4–34

(a) Angle lintel

(b) Structural tee lintel   **Figure 4–35**

This can produce a potential problem when dealing with angles or channels carrying significant loads. Although we will not labor over this issue, numerically, we will briefly discuss the nature of this problem. To this end, let's consider the behavior of a channel section as shown in Figure 4-36a. The shearing stresses throughout such a section will, in fact, be as shown in Figure 4-36b. The total effect of these shearing stresses is shown in Figure 4-36c. The shear force (V) due to the applied load is equilibrated vertically by the internal shearing resistance of the member. The shearing forces that are produced in the flanges will cause rotation about the centroid of the section. The moment produced by the forces *(H)* must be counteracted by a moment that is equal and opposite. This can only be produced by the vertical forces *(V)*. Therefore, when placing a load on an unsymmetrical member such as this, the resultant shearing force must pass through a point at a distance *e* from the centerline of the web if torsion is to be prevented. This point is known as the *shear center*. The location of the shear center is actually independent of the amount of external shear on the section. The locations of the shear center, for channel sections are given in the AISC *Manual of Steel Construction.*

When using members such as a channel or angle, where the load is large or the span is relatively long, and the load is placed so that the resultant will not pass through the shear center (which is a most common occurrence), it is recommended that these members be stabilized so as to resist the tendency to twist. Where the loads on unsymmetrical members are not very large and the spans relatively short (such as a lintel in a modestly sized window or door opening), the idea of torsion may not be a great issue.

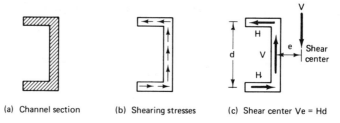

(a) Channel section   (b) Shearing stresses   (c) Shear center Ve = Hd

Figure 4-36

PROBLEMS

4.1. For each of the following, determine the maximum safe load *(W or P)* based on A36 steel and the AISC Specification. Consider that the beams are properly laterally supported.

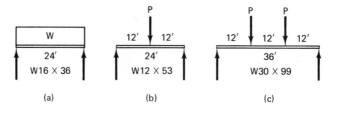

(a)   (b)   (c)

4.2. Select the most economical (least-weight) wide-flange section for each of the following (A36 steel, AISC Specification). Assume that all members will have the compression flanges laterally supported as required.

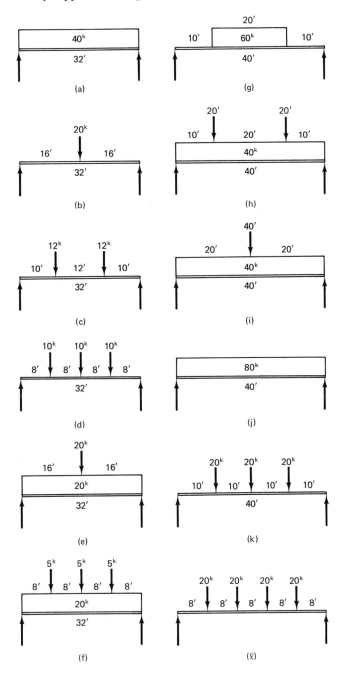

4.3. Determine the most economical wide-flange section for each of the following (A36 steel, AISC Specification). Lateral support is provided only at the concentrated loads, or under the uniform loads. Compare these to the sections required if supplementary lateral bracing is used as required.

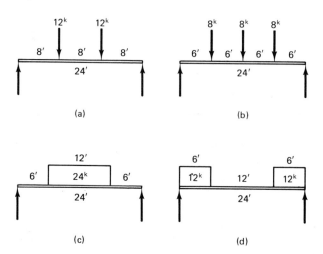

4.4. Determine the most economical wide-flange section required for each of the following cantilevered beams. Consider that the *top* flange is continuously laterally supported (A36 steel, AISC Specification).

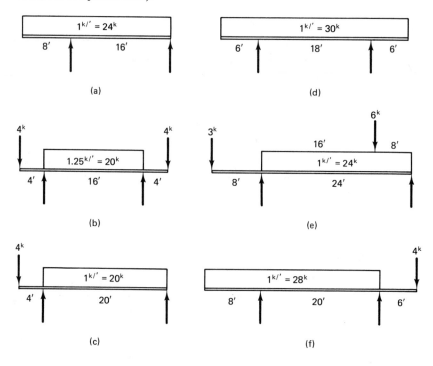

4.5. For each of the framing plans shown, determine the most economical wide-flange sections for the designated beams. Determine deflection for designated beams and, if necessary, redesign these members to satisfy deflection limitations.

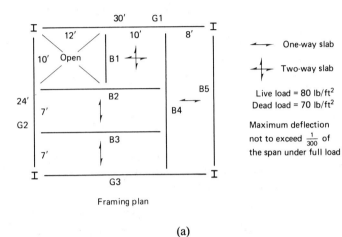

Framing plan

(a)

Framing plan

(b)

4.6. For each of the cantilevered beams:

   (1) Determine the most economical wide-flange section, using A36 steel. Assume that appropriate lateral support is provided for the compression flanges.
   (2) For beam B, determine the deflection at the midspan, under full load, and at the tips of the cantilevers.
   (3) For beam B, consider that the live load is not present and recompute the deflections.

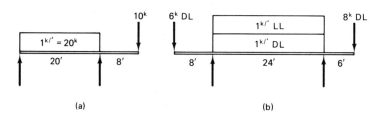

(a)   (b)

4.7. Determine the most economical wide-flange section for the continuous beams, using A36 steel. The top flange is laterally supported by the uniform load. Where only point loads are present, the top flange is laterally supported only at those points.

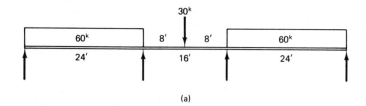

(a)

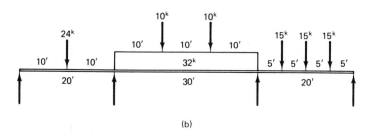

(b)

4.8. Determine the most economical wide-flange beam for each of the following based on critical arrangements of dead and live loads (A36 steel). The top flange is continuously laterally supported.

(a)   (b)

4.9. For the given framing plan, live load = 100 lb/ft²; dead load, 4-in. concrete slab = 50 lb/ft².

(a) Determine the most economical wide-flange section for B1, B2, B3 without benefit of composite action, using A36 steel.

(b) Using the members selected in part (a), determine the uniform load carrying capacity for each, with composite action. Use concrete; allowable stress = 1.8 ksi. Modular ratio $(n) = \frac{1}{8}$.

(c) Determine the deflection for B2 with and without composite action.

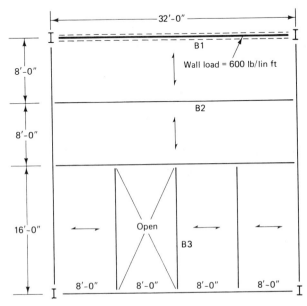

Framing plan

4.10. *Composite section.* Steel; $F_a = 24$ ksi, concrete, allowable $F = 1.35$ ksi, $n = \frac{1}{9}$.

(1) Determine the resisting moment capacity of the composite section shown.

(2) Determine the safe uniform load for a simple span of 20 ft.

(3) Determine the deflection of the composite section under this load.

(4) Determine the most economical wide flange section needed to carry the same load without composite action, and determine the deflection.

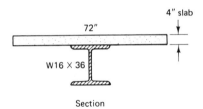

Section

4.11. *Composite section.* *Steel;* F_a = 24 ksi, concrete, allowable F = 1.8 ksi, $n = \frac{1}{8}$.
    (1) Check the safety of the composite section shown, under the given loading, by making a comparison of the actual stresses with the allowable stresses.
    (2) Determine the maximum stress in the W24 × 68 under the same loading, without composite action.

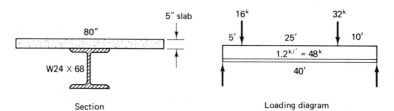

                    Section   Loading diagram

4.12. Steel; F_a = 24 ksi, concrete, allowable F = 1.35 ksi, $n = \frac{1}{9}$.
    (1) Determine the most economical wide-flange section needed without composite action, and the deflection for this condition.
    (2) Determine the effective width of the concrete flange as a composite section, using a W36 × 135 as shown.
    (3) Determine the stresses for the composite section, and the deflection.

             dead + live load = 200 lb/ft² (including slab)

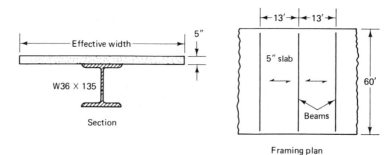

                    Section   Framing plan

TENSION AND COMPRESSION MEMBERS— ALLOWABLE STRESS DESIGN

TENSION MEMBERS

Members subjected to a pure state of tension may be found in a variety of situations in an architectural structure. They may be used as hangers for stairs, walkways, balconies, and so on (see Figure 5–1). Tension members also appear in trusses and diagonal bracing systems. There are a variety of cross sections that may be used to act in a pure state of tension, in any of the aforementioned applications. Some sections commonly used are shown in Figure 5–2. Actually, the type of cross section that may be used is limited only by the designer's imagination and needs.

The design of a tension member is a relatively simple process. The AISC Specification recommends an allowable stress on the gross cross-sectional area of a tension member not to exceed

$$F_a = 0.60F_y$$

where F_a = allowable stress
F_y = yield stress of the steel

When the member is an eye-bar with a pinhole, the allowable stress is given as $0.45F_y$ on the net section across the pinhole. We concern ourselves here only with tension members made of structural shapes, which are by far the most commonly used in an architectural structure.

In addition to the allowable stresses specified, the AISC Specification recommends minimum slenderness ratios for tension members. It should be recalled from basic studies that the *slenderness ratio* is defined as the ratio of the length in inches to the least radius of gyration *(r)* of the member. The reader should have encountered

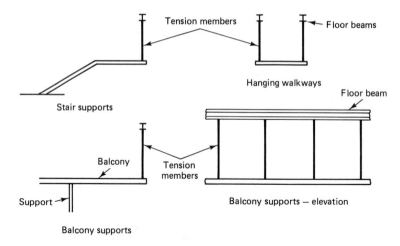

Figure 5-1

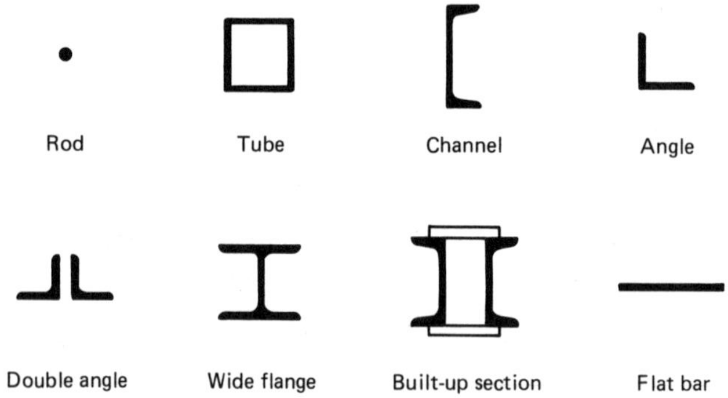

Figure 5-2 Tension members.

this expression in fundamental studies of column behavior. Column buckling is an important issue and the slenderness ratio is an important factor in determining the critical load a column may take before buckling. It may sound strange that there are recommendations for the slenderness ratio of tension members since buckling is by no means an issue. The primary purpose of such recommendations for tension members is to provide a guideline for the selection of members that will be reasonably stiff and will not vibrate excessively. The guidelines are as follows:

$$\frac{L}{r} \leq \quad 240 \quad \text{for main members}$$

$$300 \quad \text{for bracing and secondary members}$$

A main member may be defined as one whose failure would result in immediate and catastrophic collapse of a portion of the structure.

Considering that we are limiting the discussion to cross sections made of structural shapes, the only issue of great concern in the design process would be the determination of the net section. Depending on how a tension member is connected to the member being supported, this may or may not be an issue. If the connection is all welded, the net section is simply the full cross-sectional area of the member.*

If, however, the connection is made with bolts, the holes obviously reduce the cross-sectional area at the point of connection. In general, the expression "net area" refers to the cross-sectional area less any holes. For situations such as this, the AISC Specification requires that the area to be subtracted be taken as the product of the diameter of the bolt plus $\frac{1}{16}$ in., and the thickness of the material.

At this point it may be useful to present several examples for the purpose of illustrating the manner in which the preceding recommendations are used.

Example 5–1

A flat plate, 6 in. $\times \frac{1}{2}$ in., is bolted to a member with four $\frac{5}{8}$ in. round bolts, as shown in Figure 5–3. The bolts will be subjected to shear when a tensile force *(P)* is applied. Considering that the bolts can safely resist 51.6 kips, determine the tensile capacity of the plate using A36 steel. The allowable stress is 22 ksi.

Solution   The net area of the plate, considering a line across two bolt holes as the critical section, is

$$\text{net area} = 6 \times \frac{1}{2} - 2(\frac{5}{8} + \frac{1}{16})(\frac{1}{2}) = 2.31 \text{ in.}^2$$

$$P = (2.31 \text{ in.}^2)(22 \text{ ksi}) = 50.9 \text{ kips}$$

which is slightly less than the bolt capacity, and the maximum safe tensile load is, therefore, 50.9 kips.

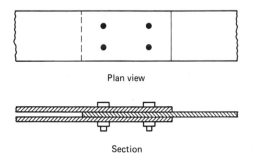

Plan view

Section   **Figure 5–3**

Example 5–2

Design a structural tee for the bottom chord of a truss, which is subjected to a tensile force of 80 kips and is supported at 10-ft intervals.

Solution   Since tees are cut from wide-flange shapes we can use Data Sheets A5 to A11 and take one-half of the area. Based on A36 steel, the allowable

---

*Be aware that, often, a temporary bolt may be used to hold the member in position while welding. The holes must be subtracted from the gross cross-sectional area.

stress is 22 ksi. Therefore, the cross-sectional area required is

$$A = \frac{80 \text{ kips}}{22 \text{ ksi}} = 3.64 \text{ in.}^2$$

One-half of a W16 × 26 will do the job. The tee section would be designated as a WT8 × 13. As discussed earlier in this chapter, the AISC requires that the slenderness ratio *(L/r)* should not exceed 240 for a main member. The bottom chord of a truss is, indeed, a main member. The least radius of gyration should be used, and this is the value of *r* with respect to the *y*-axis (the vertical axis) of the tee. The value of r_y is the same for the tee as for the wide flange from which it is cut. Therefore,

$$\frac{L}{r_y} = \frac{(10 \text{ ft})(12 \text{ in./ft})}{1.12 \text{ in.}} = 107 < 240$$

This problem is now complete.

Sag rods. Although it was mentioned previously that we will limit our concerns to members made of structural shapes, it seems that it would be useful to mention that threaded rods are often used as tension members, albeit such use is generally limited to industrial structures such as mill building types. In such structures the tension members may be sag rods supporting girts which are generally channel sections to which the outer skin of the building is attached, as shown in Figure 5-4.

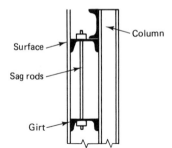

Figure 5-4

The channel sections are subjected to bending about the weak axis and, therefore, must normally be supported at several points between the columns. In addition to this, sag rods are commonly used to support roof purlins on sloping roofs where the purlin requires intermediate support because of bending about its weak axis, as shown in Figure 5-5. Since the rods used for such purposes are threaded, the net cross-sectional area of the rod is taken based on the diameter at the root of the thread. Detailed data for threaded rods may be found in the AISC *Manual of Steel Construction.*

Staggered holes. Occasionally, situations may arise whereby a tension member has holes that are staggered. The determination, in such cases, of the net cross-sectional area may not be quite obvious. For example, referring to Figure 5-6,

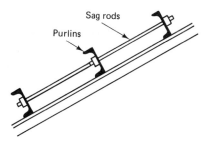

Figure 5-5

it should be clear that in part (a) the net cross-sectional area would be based on the gross area, less the width of the two bolt holes. However in part (b), where the holes are staggered, failure may occur along a zigzag pattern. For staggered holes, the determination of the least net area is a process well outlined in Section 1.14 of the AISC Specification. Should the student be interested in this sort of problem, reference should be made to this source.

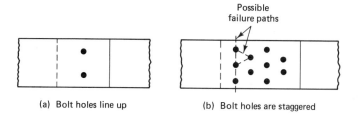

(a) Bolt holes line up   (b) Bolt holes are staggered

Figure 5-6

Eccentric connections.   Finally, in regard to tension members, it is strongly encouraged that all connections between tension members and the members they are supporting be symmetrical. For example, in Figure 5-7, two situations are shown.

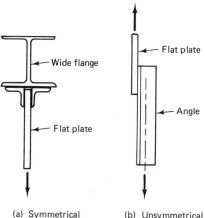

(a) Symmetrical   (b) Unsymmetrical   **Figure 5-7**

In Figure 5-7a the connection between the parts involved is symmetrical and, therefore, the axial tensile force will simply transfer from one member to the other along the centroidal axes of the members. Figure 5-7b shows a situation where the centroidal axes of the members being connected do not coincide. Because of this eccentricity at the junction, a moment will be introduced, which must be evaluated in the design process.

COMPRESSION MEMBERS

Compression members are found in a variety of situations in a building structure. We concern ourselves here with the most important and obvious compression member, that is, the column. While there are a variety of compression members, such as base plates, short piers, and so on, the distinction between a column and the others lies in the relationship between the cross-sectional dimensions and the length. It may be said that *a column is a long and slender compression member.*

In addition, there is a strong distinction between a column and a short compression member in the failure mechanism. A short compression member will crush under a load when the ultimate strength is reached. A column, which is long and slender, will buckle at some loading well below that of the ultimate strength of the material. Such a failure signals the *limit of usefulness* of the member, in spite of the fact that the material is still intact. Experience has shown that the slightest lateral force applied to a column after it has initially buckled will cause the column to fail. Therefore, the *limit of usefulness* may be defined as that load which first causes the column to buckle. This load is called the critical load, P_{cr}. If we had such a thing as an "ideal" column, with the axial load placed absolutely on the centroid of the cross section, it would be appropriate to question why the column would buckle at all. An "ideal" column is one that complies with the following:

1. The column must be absolutely straight, initially.
2. The material must be absolutely homogeneous.
3. The load must be placed absolutely concentrically.
4. The ends of the column must be absolutely aligned.

Obviously, the number of "absolutes" involved precludes the possibility of the manufacture and installation of an "ideal" column. In addition, there are normally residual stresses present in steel sections due to the manufacturing process. These residual stresses are due to rolling, rate of cooling, and the cross-sectional configuration. The cooling of a rolled section is not uniform and, consequently, there will be initial tensile and compressive stresses, not distributed uniformly. Therefore, in light of the preceding discussion, all columns will buckle under a load which is less than that indicated by the ultimate strength of the material.

End Conditions

An important influence on the load-carrying capacity of a column is the manner in which it is anchored at the ends. The type of anchorage used will have an effect on the length of the column subject to buckling which, for certain conditions, will not be the full actual length. The *effective length* of a column is the distance between the points of contraflexure of the buckled shape. The effective length may be determined by multiplying the actual length by a stiffness factor, *K,* which is based on the end conditions of the column. There are a variety of end conditions possible, but for the immediate purpose of this discussion let us consider three end condition possibilities as shown in Figure 5-8. The *K* factors shown indicate that a given column can carry the greatest load when both ends are fixed because its effective length will be only one-half of its actual length, as indicated in Figure 5-8c. The most common condition, however, in a building structure is that where both ends are pinned and, therefore, free to rotate. A condition of this sort is shown in Figure 5-8a. The connection at the foundation will normally not appreciably restrain rotation at this point. The roof beams, which are supported by the column, will keep the top of the column from moving laterally if the building as a whole is properly braced, but will not restrain rotation at this point. For a condition where both ends are pinned, the effective length and the actual length are the same, as indicated in Figure 5-8a. A third condition, where one end is fixed and the other pinned, as shown in Figure 5-8b, suggests that the effective length be taken as 0.70 times the actual length.

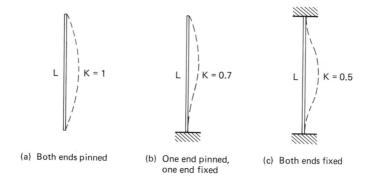

(a) Both ends pinned   (b) One end pinned,   (c) Both ends fixed
                             one end fixed

Figure 5-8   Column end conditions.

Actually, truly pinned ends or fixed ends are somewhat ideal. The "pinned" connection will normally provide a small amount of restraint against rotation and, consequently, in reality the effective length of the column may be somewhat less than the actual length. In the interest of being conservative, however, it is advisable to use a *K*-value equal to 1.0. The AISC Specification recognizes that there will be some difference between the theoretical values of K and the more realistic values, and has published recommendations of K-values to be used for various conditions. These recommendations are shown in Figure 5–9. One part of this figure is worthy of special note. Part (f) shows a situation where the column has moved laterally, at the top. An example of this sort of occurrence is shown in the frame of Figure 5–10. When a frame such as this is unbraced, sideway will occur. For a case such as this it is recommended that a K-value of 2.0 be used to provide an extra measure of stiffness for the frame.

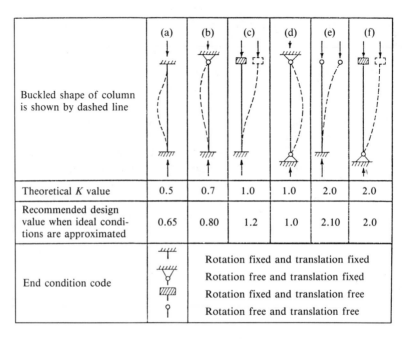

Figure 5–9   Recommended K values. From the *Manual of Steel Construction.* Reprinted with permission of The American Institute of Steel Construction.

Frames, especially of high-rise structures, can present a problem in the determination of the K-value to be used for the column design if the frame has no positive system of diagonal bracing, or shear walls, to prevent lateral translation. In our discussions we will consider that our buildings are positively braced by some means. For those concerned about multistory buildings without substantial lateral bracing, the AISC provides a means for determining the appropriate K-values to be used.*

Considering situations where columns are several stories tall, intermediate floor beams will act as lateral braces for the column, assuming that the building as a whole is properly braced against lateral displacement. The beams framing into the column

*See Commentary on the AISC Specification, Section 1.8, in the *Manual of Steel Construction,* 8th ed.

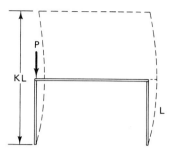

Figure 5-10 Frame sidesway.

will restrain the column from displacing laterally but will not keep the column from rotating at the point of connection. Therefore, the buckling mode will be as shown in Figure 5-11. The effective length of the column will be the length between the floor beams.

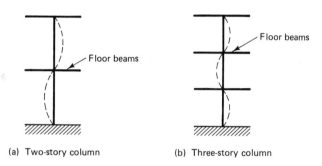

(a) Two-story column   (b) Three-story column

Figure 5-11

The Euler Formula

In the mid-eighteenth century, a Swiss mathematician by the name of Leonhard Euler performed a series of experiments to determine the buckling load for columns. These experiments led to the development of the Euler formula, which gives the load under which a long column will fail by buckling. The Euler formula has served as the basis for further research and the development of modern formulas that are used for the safe design of columns. The derivation of the Euler formula is beyond the scope or intent of this book.* We will spend our time defining and discussing the meaning of the expression. The development of the Euler formula was based on certain assumptions. They are:

1. The column is a long column and failure occurs in a buckling mode (as opposed to crushing or a combination of crushing and buckling).

2. The unit stress in the cross section at the time of failure does not exceed the elastic limit of the material.

*For those interested in the derivation of the Euler formula, see S. Timoshenko and D. H. Young, *Elements of the Strength of Materials,* 5th ed. (New York: Van Nostrand Reinhold Company, 1968).

The Euler formula, in a somewhat modified form, is

$$\frac{P_{cr}}{A} = F_{cr} = \frac{\pi^2 E}{\left(\dfrac{KL}{r}\right)^2}$$

where P_{cr} = load that first produces buckling, thereby defining the limit of usefulness
of the column
A = cross-sectional area of the column
F_{cr} = unit stress due to P_{cr}
E = modulus of elasticity of the material
KL = effective length of the column, inches (as discussed earlier)
r = radius of gyration, which is defined as

$$r = \frac{I}{A}$$

where I = moment of inertia of the cross section
A = cross-sectional area of the column

In studying the Euler formula, it should be understood that buckling will take place about the axis with the least moment of inertia, which is, consequently, the axis with the least radius of gyration. It should also be understood that the effective length has a great deal to do with the axis about which a column will buckle. For example, if a column is pinned at the ends and unbraced in either direction, it will, clearly, buckle about the axis with the least radius of gyration. If, however, the column is braced against buckling about its weak axis at some point along its length, there will be two different effective lengths involved, each related to the appropriate axis. In a case such as this it is not so clear which way the column will buckle. To determine which axis is critical where two different effective lengths and two radii of gyration are involved, we must deal with the length and radius of gyration for each axis, not as isolated values, but as a ratio, which is the *slenderness ratio, KL/r*. When comparing two axes and lengths to determine which way buckling will occur, the larger slenderness ratio will be the critical one. To clarify this issue, let us consider the following example.

Example 5–3

A W8 × 24 is to be used for a two-story column, as shown in Figure 5–12. It is laterally braced in the weak direction at the floor level, but unbraced in the strong direction. Determine the critical slenderness ratio.

Solution The critical length for the y-axis is 12 ft. Therefore,

$$\frac{KL_y}{r_y} = \frac{(12 \text{ ft})(12 \text{ in./ft})}{1.61 \text{ in.}} = 89.4$$

Since the column is unbraced in the strong direction, then

$$\frac{KL_x}{r_x} = \frac{(20 \text{ ft})(12 \text{ in./ft})}{3.42 \text{ in.}} = 70.2$$

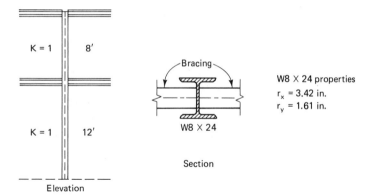

Figure 5–12

The larger of the two slenderness ratios will be critical and the maximum load-carrying capacity will be controlled by the 12-ft length and the *y*-axis.

AISC FORMULAS FOR DESIGN OF COMPRESSION MEMBERS

It should be clear that the Euler formula provides us with information about the load (P_{cr}) or the unit stress (F_{cr}) at which buckling will occur, under ideal conditions. We would not actually want to design a column in a building to buckle under the applied load. Consequently, factors of safety must be applied in order to determine the allowable stress *(F_a)* to be used in the column design process. Such formulas are known as "working" or "design" formulas. There are a variety of design formulas available, depending on the specification being used. We concern ourselves only with the AISC Specification.

The Euler formula has served as a basis for a great deal of research and experimentation. Through the years the AISC has developed somewhat more accurate and meaningful formulas for design which are related to the Euler formula. Essentially, the AISC formulas take into account several issues not considered by the pure form of the Euler formula, such as the presence of residual stresses and true (not idealized) end conditions of a column. The AISC formulas are, to some extent, empirical. Consequently, any attempt at a mathematical derivation would be very complex and somewhat meaningless within the scope or intent of this book. Suffice it to say that based on the AISC formulas, the allowable stresses for various slenderness ratios have been computed and are included in the Appendix. We would do better to use our time on an example problem using the already computed values provided by the AISC.* The computed values of the allowable stress for various values of slender-

---

*For those interested in seeing the AISC Formulas for the design of compression members, see *Manual of Steel Construction,* 8th ed., AISC Specification, Section 1.5.

ness ratios shown in the Appendix (Data Sheet A24) are based on ASTM A36 steel $(F_y = 36$ ksi).*

Example 5–4

Select a wide-flange column to carry an axial load *(P)* of 150 kips. The column is 12 ft long, has pinned ends, and is unbraced. Use A36 steel and the AISC Specification. (See Data Sheet A24 for allowable stresses.)

Solution   The procedure for selecting a column section is one of trial and error, since we do not know a value of the radius of gyration *(r)* at the outset. Using the allowable stress table in the Appendix, let us start out by assuming $F_a = 18$ ksi.†

$$\text{required area } (A) = \frac{150 \text{ kips}}{18 \text{ ksi}} = 8.33 \text{ in.}^2$$

Scan the wide-flange properties tables (Data Sheets A5 to A11):

Trial 1: W8 × 28, $A = 8.25$ in.^{2}, $r_y = 1.62$ in.

$$\frac{KL}{r_y} = \frac{(12 \text{ ft})(12 \text{ in./ft})}{1.62 \text{ in.)}} = 88.9 \qquad F_a = 14.32 \text{ ksi} \qquad \text{(no good)}$$

This indicates that we need a larger member. The first trial run gives a basis for rapidly closing in on the proper selection. We need a section with a little more *r* in order to increase the allowable stress, and a little more cross-sectional area.

Trial 2: W8 × 31, $A = 9.13$ in.^{2}, $r_y = 2.02$ in.

$$\frac{KL}{r_y} = \frac{144 \text{ in.}}{2.02 \text{ in.}} = 71.3 \qquad F_a = 16.3 \text{ ksi}$$
$$P = (16.3 \text{ ksi})(9.13 \text{ in.}^2) = 148.8 \text{ kips}$$

This is very close to the required capacity of 150 kips. It should be clear from scanning the properties tables that there are only two sensible choices left for trial as the most economical section: the W10 × 33 and the W8 × 35. Let us try the W10 × 33 since it is a bit lighter.

Trial 3: W10 × 33, $A = 9.71$ in.^{2}, $r_y = 1.94$ in.

$$\frac{KL}{r_y} = \frac{144 \text{ in.}}{1.94 \text{ in.}} = 74.2 \qquad F_a = 16 \text{ ksi}$$
$$P = (16 \text{ ksi})(9.71 \text{ in.}^2) = 155.4 \text{ kips}$$

The W10 × 33 is the most economical section, and the design problem is complete.

---

*Computed values of allowable stress for steel with $F_y = 50$ ksi may also be found in the *Manual of Steel Construction*, 8th ed., AISC Specification, Appendix A.

†Using A36 steel the allowable stress in a compression member will always be less than 22 ksi, and generally in the "teens."

In Example 5-4 the second trial selection (W8 × 31) was very close to satisfying the load requirement. Actually, the difference is less than 1%. Under normal circumstances this would be an appropriate selection. The decision to do this must rest with the designer and may be based on judgments about the actual load that may be experienced, the number of columns (and the consequent weight savings), and so on. On the other hand, it should be realized that a column is one of the most critical structural elements in a building and failure could be catastrophic. Therefore, as always, sound judgment plays a key role in such decisions.

While we tend to take the word "column" as meaning a vertically oriented compression member, we often encounter compression members that are not obvious as columns, but whose behavior is precisely the same. A prime example of this would be the compression members in a truss, as shown in Figure 5-13. Depending on the

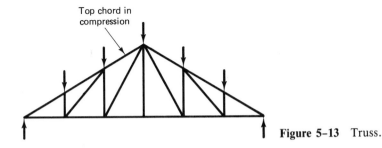

Top chord in
compression

Figure 5-13   Truss.

configuration of the truss, a variety of web members will be subjected to pure states of axial compression, and assuming gravity loadings, the top chord will also be a compression member, provided that the truss is not cantilevered. The behavior of these members subjected to axial compression would be precisely the same as a vertically oriented column and their design would follow the requirements of the AISC Specifications. Normally, in a steel structure, the members of which a truss is made are double-angle members, as shown in Figure 5-14. The AISC *Manual of Steel Construction* provides complete load tables for the great variety of double-angle combinations that are possible and the design of such members may be expedited by referring to the *Manual*.

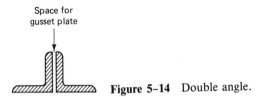

Space for
gusset plate

Figure 5-14   Double angle.

COMBINED BENDING AND AXIAL STRESS

It is somewhat of an idealistic situation to have a column that is subjected to pure axial stress. There will normally be some bending moment involved in addition to the axial load. There are several reasons for this: it is virtually impossible to center loads exactly on the centroidal axes of the column cross section, the column is not initially straight, and the column may be subject to transverse loading. In addition, beams are normally detailed to frame into the side of a column, as shown in Figure 5–15, as opposed to resting on top of the column. This is normally even true where the column does not extend upward, as in a roof. The beam reaction being delivered to a point that is eccentric to the centroidal axis of the column produces a bending moment. If the column is very small, with a consequent small eccentricity, the bending moment is probably negligible. When the column has large amounts of both axial and bending stresses, it is often referred to as a *beam-column*.

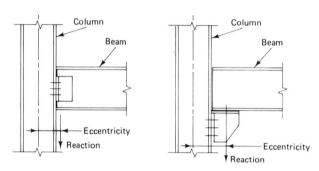

Figure 5–15

When dealing with a member subjected to both axial and bending stresses, it may be recalled, from structural basics, that the stress *(f)* at any point within the cross section may be obtained from the *interaction equation,* which is as follows:

$$f = \frac{P}{A} \pm \frac{Mc}{I}$$

Although this is the fundamental expression used to determine the magnitude of combined stresses, it does not make any allowance for the effect of increased deflections caused by the bending moment, and the consequent increased bending moment due to the greater eccentricity of the compressive force (see Figure 5–16).

The AISC Specification permits the use of the basic interaction equation up to a certain point. The expression suggested is

$$\frac{f_a}{F_a} + \frac{f_b}{F_b} \le 1.0 \tag{1}$$

where f = computed (actual) axial stress, ksi
F_a = axial compressive stress permitted if only axial load existed, ksi
f_b = computed compressive bending stress at point being investigated, ksi
F_b = compressive bending stress permitted if bending moment alone existed, ksi

If bending took place about both axes, which is quite possible in a corner column, we would include bending expressions for each axis and the expressions would be as follows:

$$\frac{f_a}{F_a} + \frac{f_{bx}}{F_{bx}} + \frac{f_{by}}{F_{by}} \leq 1.0 \qquad \text{[AISC Eq. 1.6-2]} \qquad (2)$$

The validity of these equations is limited, by the AISC Specification, to conditions where the axial stress accounts for 0.15 (or less) of the available strength of the selected member. In other words, this expression may be used when conditions are such that the member is more of a beam than it is a column.

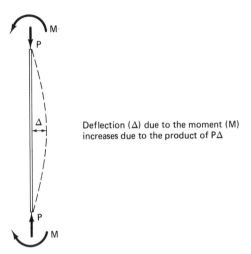

Deflection (Δ) due to the moment (M) increases due to the product of PΔ

Figure 5–16

When the axial stress becomes large, relative to bending stresses, the AISC requires a more conservative approach to account for increased moments due to the eccentricity created by lateral deflection. For these conditions there are two expressions that must be checked. The AISC equations are

$$\frac{f_a}{F_a} + \frac{C_{mx}f_{bx}}{\left(1 - \dfrac{f_a}{F'_{ex}}\right)F_{bx}} + \frac{C_{my}f_{by}}{\left(1 - \dfrac{f_a}{F'_{ey}}\right)F_{by}} \leq 1.0 \qquad \text{[AISC Eq. 1.6-1a]} \qquad (3)$$

$$\frac{f_a}{0.60F_y} + \frac{f_{bx}}{F_{bx}} + \frac{f_{by}}{F_{by}} \leq 1.0 \qquad \text{[AISC Eq. 1.6-1b]} \qquad (4)$$

The subscripts x and y in these expressions refer to the axes of the cross section. If bending exists about only one axis, one of the expressions in Equations 3 and 4 will drop out. The expressions f_a, F_a, f_b, and F_b are the same as previously defined and

$$F'_e = \frac{12\pi^2 E}{23(Kl_b/r_b)^2} \qquad (5)$$

which is the Euler formula (in ksi) with a factor of safety of 12/23 applied. L_b is the unbraced length (inches) in the plane of bending and r_b is the corresponding radius of gyration. C_m is a factor that varies with loading and restraint conditions.

According to the AISC Specification there are, basically, three categories to be considered.* They are:

1. $C_m = 0.85$ for members in frames subjected to sidesway (no positive bracing system).

2. Where compression members are restrained, the frame is braced against sidesway, and there is no transverse loading between column supports:

$$C_m = 0.6 - 0.4 \frac{M_1}{M_2} \qquad \text{but not less than 0.4}$$

where M_1/M_2 is the ratio of the smaller to the larger moments at the restrained ends. This ratio is positive when the member is bent in reverse (double) curvature and negative when bent in single curvature (see Figure 5–17).

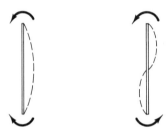

    Single curvature   Reverse (double) curvature **Figure 5–17**

3. For members in frames braced against sidesway in the plane of loading and subjected to transverse loading between supports, the following values may be used:

Members with restrained ends: $C_m = 0.85$
Members with unrestrained ends: $C_m = 1.0$

In Equation 3, the expression

$$\frac{1}{1 - \dfrac{f_a}{F'_e}}$$

is called the *amplification factor,* which is applied to the basic form of the interaction equation to account for the additional moments due to lateral deflections. The factor, C_m, is a reduction factor that keeps the amplification factor from being too large (and, hence, overly conservative), which can happen due to a variety of conditions. Considering the function of the factor, C_m, it should be clear that its value will never exceed 1.0.

At this point it seems that it would be useful to apply the AISC interaction equations in the context of a specific example problem. For this purpose a member was selected from a *braced* building frame.

---

*For a complete discussion, and rationale, see Section 1.6 of the Commentary on the AISC Specification in the *Manual of Steel Construction,* 8th ed.

Example 5-5

Design a wide-flange column to support the axial load and bending moment shown in Figure 5-18. The frame is braced against sidesway, and the column is not subject to transverse loading. Use A36 steel and the AISC Specification. Use $K = 1$.

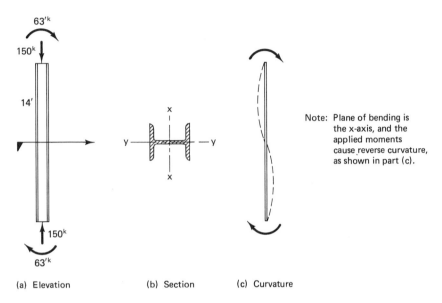

63'k

150k

14'

x

y ——— y

x

150k

63'k

Note: Plane of bending is the x-axis, and the applied moments cause reverse curvature, as shown in part (c).

(a) Elevation   (b) Section   (c) Curvature

Figure 5-18

Solution   The column design process, as suggested in Example 5-4, is one of trial and error. In this problem we have the additional complexity of bending moments producing a combined stress situation. To determine a section to be used as a first trial it is suggested that a section be selected based on a low level of allowable axial stress (F_a). In this case let us say that $F_a = 10$ ksi. Considering the axial load shown, this would suggest a section with a cross-sectional area of about 15 in.^{2}. Therefore, scanning the wide-flange properties tables (Data Sheets A5 to A11), let us try a W10 × 54.

Trial 1: W10 × 54, $A = 15.8$ in.^{2}, $r_y = 2.56$ in., $r_x = 4.37$ in. We must first check the unbraced length tables (Data Sheets A21 to A23) to determine the appropriate bending stress to be used, considering $KL = 14$ ft:

For a W10 × 54; $L_c = 10.6$ ft, $L_u = 28.3$ ft

Therefore, $F_b = 22$ ksi (noncompact section). Since the member is unbraced, the y-axis is critical for axial stress, and

$$\frac{KL}{r_y} = \frac{(14 \text{ ft})(12 \text{ in./ft})}{2.56 \text{ in.}} = 65.6 \qquad F_a = 16.88 \text{ ksi (allowable)}$$

$$f_a \text{ (actual)} = \frac{P}{A} = \frac{150 \text{ kips}}{15.8 \text{ in.}^2} = 9.5 \text{ ksi}$$

Test Equation 2 (bending about x-axis only)

$$\frac{f_a}{F_a} = \frac{9.5 \text{ ksi}}{16.88 \text{ ksi}} = 0.56 > 0.15$$

This suggests that we must evaluate the trial section using Equations 3 and 4 (AISC Eqs. 1.6–1a and 1.6–1b). For columns restrained against sidesway and in reverse curvature (this is our case), Equation 4 will normally be critical. Since this equation is relatively simple to evaluate, let us deal with it first and then we can decide whether to continue or to go to another trial section.

$$(4) \quad \frac{f_a}{0.60F_y} + \frac{f_{bx}}{F_{bx}} \leq 1.0$$

The only component of this equation that has not yet been evaluated is f_{bx} (the actual bending stress). For a W10 × 54, $S_x = 60 \text{ in.}^3$.

$$f_{bx} = \frac{M}{S} = \frac{(63 \text{ ft-kips})(12 \text{ in./ft})}{60 \text{ in.}^3} = 12.6 \text{ ksi}$$

Therefore (going to Equation 4),

$$\frac{9.5 \text{ ksi}}{22 \text{ ksi}} + \frac{12.6 \text{ ksi}}{22 \text{ ksi}} = 1.00 \qquad \text{(right on!)}$$

We cannot do much better than this, but we still should check the results of Equation 3. Since there is only one bending axis,

$$\frac{f_a}{F_a} + \frac{C_{mx} f_{bx}}{\left(1 - \dfrac{f_a}{F'_{ex}}\right) F_{bx}} \leq 1.0$$

Let us first evaluate the factor C_m. We are dealing with a condition where the frame is braced against sidesway and the column is not subjected to transverse loading. Therefore,

$$C_m = 0.6 - 0.4 \frac{M_1}{M_2} \quad \text{(but not less than 0.4)}$$

The value of the ratio M_1/M_2 is positive since the column is bent in reverse curvature. Therefore,

$$C_m = 0.6 - 0.4 \left(\frac{63 \text{ ft-kips}}{63 \text{ ft-kips}}\right) = 0.2 \qquad \text{use } C_m = 0.4$$

We must now evaluate F'_e (Equation 5).

$$F'_e = \frac{12\pi^2 E}{23\left(\dfrac{KL_b}{r_b}\right)^2} = \frac{12(\pi)^2(29 \times 10^3)}{23\left(\dfrac{168}{4.37}\right)^2} = 101 \text{ ksi}$$

and

$$1 - \frac{f_a}{F'_{ex}} = 1 - \frac{9.5 \text{ ksi}}{101 \text{ ksi}} = 0.91$$

Using these values in Equation 3 yields

$$\frac{9.5 \text{ ksi}}{16.88 \text{ ksi}} + \frac{(0.4)(12.6 \text{ ksi})}{(0.91)(22 \text{ ksi})} = 0.82 < 1.0$$

Considering the good luck we had with the first trial section, a W10 × 54 would be a most appropriate choice. For the sake of demonstration, however, let us see if we can find a slightly more economical section. Based on the first trial, there is not too much room for selection of a second trial section. In scanning the properties tables we see that a W10 × 49 has about the same value for r_y but a smaller value for S_x. Therefore, it should be clear that it will not satisfy Equation 4. Looking at the 12-in. wide-flange sections, it seems that, perhaps, a W12 × 50 may work. Let us try it.

Trial 2: W12 × 50: $A = 14.7$ in.^{2}, $r_y = 1.96$ in., $r_x = 5.18$ in., $S_x = 64.7$ in.^{3}. Going to Equation 4:

$$f_a = \frac{P}{A} = \frac{150 \text{ kips}}{14.7 \text{ in.}^2} = 10.2 \text{ ksi}$$

$$f_{bx} = \frac{M}{S} = \frac{(63 \text{ ft-kips})(12 \text{ in./ft})}{64.7 \text{ in.}^3} = 11.68 \text{ ksi}$$

and

$$\frac{10.2 \text{ ksi}}{22 \text{ ksi}} + \frac{11.68 \text{ ksi}}{22 \text{ ksi}} = 0.99 < 1.0 \qquad \text{O.K.}$$

Considering that Equation 4 is normally critical for conditions such as ours, the numbers indicate that a W12 × 50 will work. Before going through the tedium of Equation 3, let us see if we can find a more economical section. In scanning the properties tables it seems that the only remaining possibility would be a W14 × 48.* Let us try this with Equation 4 and then make a final decision.

W14 × 48: $A = 14.1$ in.^{2}, $r_y = 1.91$ in., $r_x = 5.85$ in. $S_x = 70.3$ in.^{3}.

$$f_a = \frac{P}{A} = \frac{150 \text{ kips}}{14.1 \text{ in.}^2} = 10.64 \text{ ksi}$$

*Initially, it appeared that a W16 × 45 would work. However, a check of the unbraced lengths tables indicated a value of only 11.4 ft for L_u. It is recommended that L_u not be exceeded.

$$f_{bx} = \frac{M}{S} = \frac{(63 \text{ ft-kips})(12 \text{ in./ft})}{70.3 \text{ in.}^3} = 10.75 \text{ ksi}$$

and

$$\frac{10.64 \text{ ksi}}{22 \text{ ksi}} + \frac{10.75 \text{ ksi}}{22 \text{ ksi}} = 0.97 \qquad \text{O.K.}$$

Let us now test Equation 3. $C_m = 0.4$ (as previously determined).

$$F_e' = \frac{12\pi^2 E}{23 \left(\dfrac{KL_b}{r_b}\right)^2} = \frac{12(\pi)^2(29 \times 10^3 \text{ ksi})}{23 \left(\dfrac{168 \text{ in.}}{5.85 \text{ in.}}\right)^2} = 181 \text{ ksi}$$

and

$$1 - \frac{f_a}{F_{ex}'} = 1 - \frac{10.64 \text{ ksi}}{181 \text{ ksi}} = 0.94$$

We must now determine the value of F_a for the W14 × 48.

$$\frac{KL}{r_y} = \frac{168 \text{ in.}}{1.91 \text{ in.}} = 88 \qquad F_a = 14.44 \text{ ksi}$$

Using these values in Equation 3 yields

$$\frac{10.64 \text{ ksi}}{14.44 \text{ ksi}} + \frac{(0.4)(10.75 \text{ ksi})}{(0.94)(22 \text{ ksi})} = 0.94 \qquad \text{O.K.}$$

The most economical wide-flange section is a W14 × 48, and the design problem is complete.

Example 5-5 represents one set of conditions out of a possible large variety of conditions. It is, perhaps, somewhat of an oversimplification in terms of today's building practices. The example assumed that there was a positive system of bracing which would eliminate sidesway. With the lightweight structures we build today, it is likely that there would be no positive system of bracing, and the frame itself would have to provide the necessary stiffness. The AISC Specification provides a rational method for determining the effective lengths of columns in a frame subjected to sidesway.*

The example is solely for the purpose of providing the student with a sense of the issues and language involved in the design of a column subjected to both axial and bending stresses. In this spirit it should also be mentioned that for those who truly wish to engage themselves in the actual design of a steel column, much of the tedium suggested by the example problem would be eliminated by the use of the examples and tables provided in Section 3 of the *Manual of Steel Construction* of the AISC.

*See Section 1.8 of the Commentary on the AISC Specification in the *Manual of Steel Construction,* 8th ed.

PROBLEMS

5.1. Determine the maximum allowable axial compressive load for each of the wide-flange sections indicated. The columns are pinned at the ends and are unbraced in either direction (A36 steel, AISC Specification).

Section	KL	Section	KL
W6 × 20	10'	W10 × 22	14'
W6 × 20	15'	W10 × 49	20'
W8 × 21	12'	W12 × 22	14'
W8 × 21	16'	W12 × 50	20'
W8 × 40	12'	W14 × 82	30'

5.2. Using A36 steel and the AISC Specification, select the most economical wide-flange sections to carry the given axial compressive loads.
(1) $P = 80$ kips, $KL = 25$ ft
(2) $P = 260$ kips, $KL = 16$ ft
(3) $P = 160$ kips, $KL(x\text{-axis}) = 20$ ft, $KL(y\text{-axis}) = 12$ ft

5.3. Design of two-story column (A36 steel, AISC Specification). The two-story column shown is made of one piece. It is braced in the weak direction at the first-floor level and unbraced in the strong direction for the full height. Load at the roof = 60 kips; no load at the first floor. Consider that the column is pinned at the ends and that the building frame, as a whole, is braced (no sidesway). Determine the most economical wide-flange section.

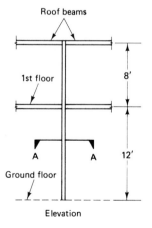

Elevation

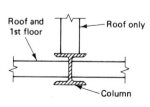

Section A-A

5.4. The section shown is to be used as a column. It is 24 ft long, with pinned ends, and it is unbraced in either direction. Using A36 steel and the AISC Specification, determine the maximum safe axial load.

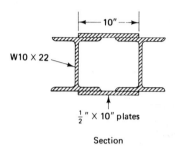

Section

5.5. The steel tubular column is 4 in. × 12 in. (outside) with $\frac{1}{2}$ in. thick walls. The ends are pinned and it is braced in the weak direction, as shown. Using A36 steel and the AISC Specification, determine the maximum safe axial load *P*.

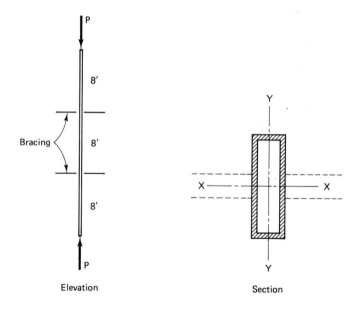

Elevation   Section

5.6. Using A36 steel and the AISC Specification, design a wide-flange steel column. The column is to be one continuous piece. Total load at the roof = 100 kips; total load at the first floor = 200 kips.

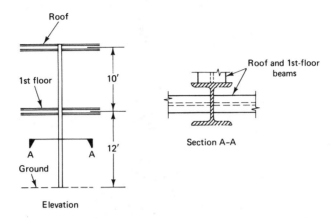

5.7. *Column investigation and design (A36 steel, AISC Specification)*

(1) W8 × 31 columns are used as main members with bracing of the weak axis, as shown. Determine the maximum safe axial load *(P)*.

(2) Determine the safe axial load if the bracing is omitted.

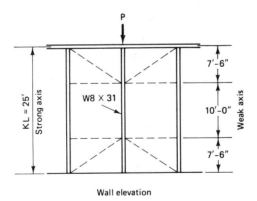

Wall elevation

5.8. *Column design (A36 steel, AISC Specification)*

A 14-ft column in a building carries an axial load of 220 kips and a moment at one end of 60 ft-kips. The moment is applied about the strong axis. The column is pinned at the ends and the building is braced against sidesway. Determine the most economical wide flange section.

5.9. *Column design (A36 steel, AISC Specification)*
Determine the most economical wide-flange section to carry the axial load and transverse loading shown. The column is pinned at the ends and there is no sidesway.

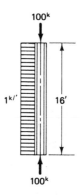

5.10. *Column design (A36 steel, AISC Specification)*
Determine the most economical wide-flange column for the condition shown. The moments are applied at the ends and are in the strong direction. The manner in which the moments are applied causes reverse curvature. There is no transverse loading and no sidesway.

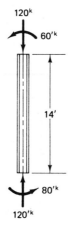

6

Joinery in Structural Steel

The design of connections between structural steel members must include consideration of all of the types of stress that the connection will be required to resist. In addition, a great deal of judgment is required in the selection of the fastening method, that is, bolted, welded, or a combination of the two. Rivets are, for all practical purposes, no longer used in building structures. The actual design of connections between framing members is best left to the engineer. The design of such connections can be somewhat complex and certainly a bit tedious. Consequently, rather than concern ourselves with numerical design procedures, we will limit our concerns here to discussions of various types of connections and the stresses to which they will be subjected. Hopefully, the student will, by concentrating on the language of connections in steel, develop a vocabulary that will be useful for the purpose of communicating with the consulting engineer and the steel fabricator.

TYPES OF CONSTRUCTION FOR STRUCTURAL STEEL FRAMES

The AISC Specification broadly outlines three types of construction for structural steel frames. We will now proceed to define the three types and elaborate on their meanings.

Type 1 construction: This classification covers ''rigid frame'' construction. In this type of construction the beam-to-column connections are rigid enough to hold the original angle between the members when the frame deforms under the applied loads. This means that the beam-to-column connection has the ability to resist the bending moment. A typical example illustrating the idea of the unchanged angle between beam and column is shown in Figure 6-1. Note that the frame is shown in its deformed shape (greatly exaggerated) but the original angle between members remains at 90°, albeit the entire junction between the beam and column has rotated. Several possible connection configurations that would classify as "rigid" are shown in Figure 6-2.

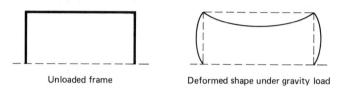

Unloaded frame   Deformed shape under gravity load

Figure 6-1

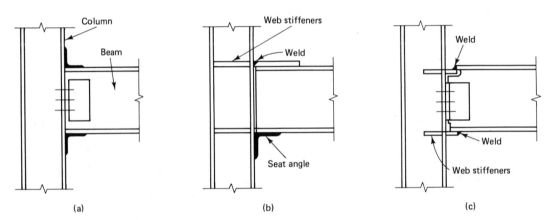

(a)   (b)   (c)

Figure 6-2   Several types of rigid connections.

Type 2 construction: This classification covers "simple" framing. The concept of the "simply supported beam" should be familiar from basic studies. This means that the ends of the beam are free to rotate relative to the support, under gravity loading (see Figure 6-3). The original angle between the horizontal and vertical

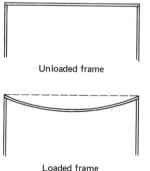

Unloaded frame

Loaded frame   **Figure 6-3**

members will, in fact, change when the beam deforms due to a load. A connection such as this (flexible as opposed to rigid) is considered to resist shear only, and cannot resist bending moment. However, it is permissible, in type 2 framing, to design the connections to resist moments due to wind. Essentially, this means that the connection must have the ability to resist shear and moment. When using a connection of this sort, the girder must be designed to carry the full gravity load as a "simply supported beam" and advantage cannot be taken of the reduction of positive moment. We discuss this sort of condition a bit further in Chapter 7. Several examples of commonly used simple connections are shown in Figure 6-4.

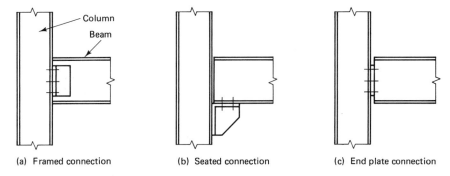

(a) Framed connection   (b) Seated connection   (c) End plate connection

Figure 6-4 Several simple connections.

Type 3 construction: This classification refers to "semi-rigid" framing. This means that the connection is designed as one that will offer partial restraint. The connection has a certain known moment capacity somewhere between the moment required for a fully rigid connection (type 1 construction) and a simple connection (type 2 construction). A use of semi-rigid connections would be in a situation where a connection must be designed to carry moments induced by wind, as well as gravity loading. This is similar to type 2 framing with wind connections except that the girder may be proportioned taking advantage of a predictable degree of full end restraint under gravity load. The AISC Specification requires that documentation be provided to show the degree of full end restraint to be used under gravity loading. Type 3 construction is uncommon in building structures. Some examples of semi-rigid connections are shown in Figure 6-5.

Fundamentally, there may be no difference in the arrangement of the parts in type 1 and type 3 connections. The differences have to do with the philosophy under which the structural designer is operating. In a type 1 connection there will be bending moments at the supports as well as shear, all due to gravity loading. In addition, the type 1 connection may be required to carry additional moments created by lateral loading. The type 1 connection would be designed to develop the end moment capacity of the beam or girder required for full rigidity. In type 3 construction the connections are also designed to resist moment as well as shear. However, the moments designed for in semi-rigid construction may be only an assumed percentage of the

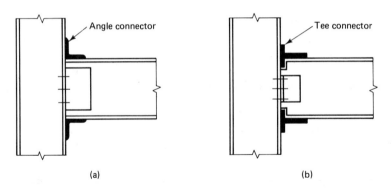

(a)   (b)

Figure 6-5   Semi-rigid connections.

moment capacity of the connected members. Some of the moment to be resisted may be due to wind.

There are also similarities between type 3 framing and type 2 framing with wind connections. For the sake of clarification, let's consider a semi-rigid connection with an assumed percentage of moment, under gravity loading, equal to zero, and the entire moment to be resisted is due to wind. This situation would adequately describe type 2 framing with wind connections. Philosophically, this connection may be thought of as one where, under the influence of gravity loading, only shear is resisted, and this is done solely by the web angles. When wind is acting on the structure and the frame tends to distort and move laterally, the resistance to this displacement is provided by the moment-resisting part of the connection. This part of the connection would be thought of as doing nothing when lateral forces are not present.

In the preceding discussion, the connections shown were based on framing systems that are rectangular. In other words, the members being connected are at right angles to each other. There are, however, many occasions when members being connected do not intersect each other at 90° angles. This may occur in buildings with nonrectangular bay patterns, sloping roofs, or a variety of other conditions. Such connections may be skewed, sloped, canted, or a combination of these. With reference to Figure 6-6, these conditions may be defined as follows:

a. When a beam or girder spanning horizontally frames into a supporting member (column or girder) and the intersection between the two is not perpendicular, the beam or girder is said to be *skewed*. This sort of condition is shown in Figure 6-6a.

b. When the web of a beam or girder is perpendicular to the supporting member but the flanges are not, the beam is said to be *sloped*. This is shown in Figure 6-6b.

c. When a beam is perpendicular to the face of a supporting member but rotated so that its flanges are tilted with respect to those of the support, as shown in Figure 6-6c, the member is said to be *canted*.

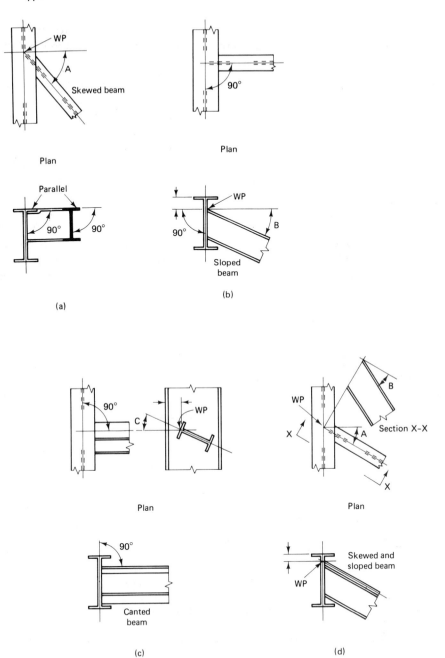

Figure 6-6 Types of connections. From *Engineering for Steel Construction.* Used with permission of The American Institute of Steel Construction.

d. When a beam is included in two directions with respect to the axis of the supporting member, the beam is *skewed and sloped*. This is shown in Figure 6–6d.*

Skewed, sloped, or canted connections are not necessarily difficult to detail or execute. However, since they are not standard, the cost for such work will probably be higher than normal. When connections are skewed and sloped, these can be quite geometrically complex and the detailing and execution may be quite labor intensive and, therefore, costly. Such conditions should be avoided where cost is a factor.

In addition to connections between beams, girders, and columns there are a variety of other connection types that will normally be encountered in a steel structure. For example, connections of bracing to a structural frame may take on a variety of configurations, depending on the actual member being used for bracing, and the design concept employed. Bracing may be concentric, such as X-bracing, or eccentric, such as K-bracing. Types of bracing are discussed further in Chapter 7. For our immediate purposes, suffice it to say that if the design philosophy calls for concentric bracing, the centroidal axis of the brace must intersect with the intersection of the centroidal axis of the girder and column. In this type of situation the force in the brace will not introduce bending moments to the columns or girders. Several possibilities are illustrated in Figure 6–7.

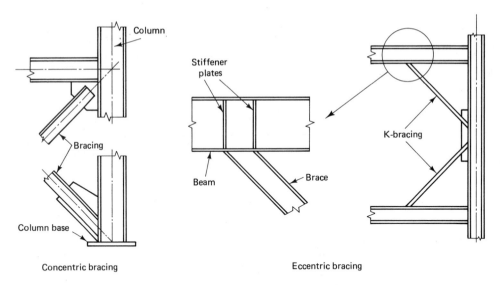

Concentric bracing   Eccentric bracing

Figure 6–7   Bracing connections.

Where a trussed structure is being used, the connection between the web members and the chord members must be considered. Normally, in a steel truss, the web and chord members are made of double-angle sections. The common method of connec-

---

*For further discussion and detailing suggestions, see *Engineering for Steel Construction* (Chicago: American Institute of Steel Construction, 1984), Chapter 5.

tion is to insert a gusset plate in a space between the double angles and welding the connection (see Figure 6-8).

There are a variety of other connections that may also be encountered. These include connections for hangers, haunches, column splices, column bases, and so on. Some possibilities are illustrated in Figure 6-9.

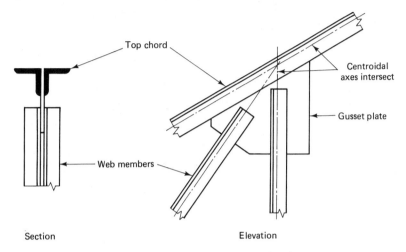

Figure 6-8   Connection at truss panel point.

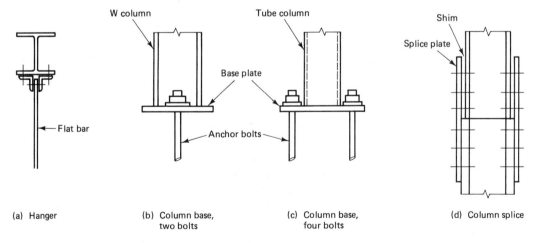

Figure 6-9

States of Stress

Considering the large variety of connection types and the possible variations of each, what should be understood about all connections are the types of stresses that must be transferred, and something about the behavior of the connection. With this kind of understanding, sensible connection geometry can be developed.

In simple beam connections to columns or girders, such as shown in Figure 6–4, it should be understood that since no moment-carrying capacity is considered to be involved, the only stresses to be transferred are shearing stresses. In a case such as this the shearing force in the connection must be, at least, equal to the end reaction of the beam. Actually, simple connections such as this may be selected from tables provided in the AISC *Manual of Steel Construction.* Because these are considered to be standard connections, the structural engineer need not even show the shearing force to be resisted by the connection. If the end reactions of the simply supported beams are not indicated on the structural drawings the fabricator's engineer will automatically select a connection to support one-half of the uniform load-carrying capacity of the member shown, as a simply supported beam. Values for the uniform load capacity of simply supported wide-flange beams for a variety of spans are tabulated and given in the *Manual of Steel Construction.* The large majority of the beams in architectural structures will fall in this category. However, it must be realized that, where special conditions exist, the structural engineer must show the actual end reactions of the beam to be supported by the connections. Special conditions may occur where beams are spanning short distances and carrying very heavy concentrated loads, or concentrated loads placed close to a support. Where composite construction is being used the reactions to be resisted by the connections must be shown on the structural engineering drawings. This is also true where beams are continuous rather than simply supported.

In rigid (type 1) or semi-rigid (type 3) construction, the connections, shown in Figure 6–2 and 6–5, must be designed to transfer not only the shearing force, but also the bending moment for which the connection is designed. The connection at the web of the beam will transfer the shear force. The top and bottom pieces combine to provide the resisting moment capacity of the connection. The top piece will be in a state of tension and the bottom piece will be in a state of compression, thereby forming a couple to provide the moment resistance.*

When considering the geometry of a connection, there may be combined stresses in the fastening system that must be considered. This can happen in a variety of situations, especially where one of the connected members is diagonally oriented. For example, let us consider the connection of a diagonal bracing member to a column, shown in Figure 6–10. If the brace is in a state of direct tension, the stress across the gusset plate will be tensile. If the plate is welded to the column, the weld will be subjected to a shear force, which is the vertical component of the force in the brace, as well as a tensile force across the weld, which is the horizontal component. If the connection is made with bolts, the bolts will be subjected to tension and *single shear.* This means that the shearing force across the bolts will be resisted by one cross section of the bolts. There are conditions where bolts may be subjected to *double shear.* Let us consider the idea of a single and double shear a bit further. If a beam is connected to a column by means of a simple connection, and the fasteners used are bolts, as shown in Figure 6–11, the bolts will be in a state of shear. The shearing

---

*If the moment is due to wind only, the location of tensile and compressive forces in the connection may be reversed, depending on the location of the connection.

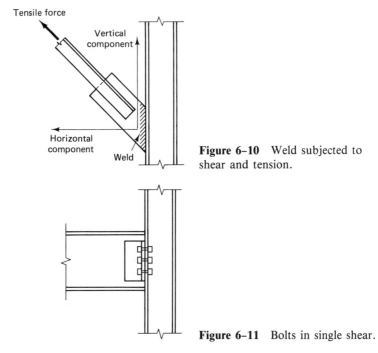

Figure 6-10 Weld subjected to shear and tension.

Figure 6-11 Bolts in single shear.

resistance provided by the group of bolts must be at least equal to the end reaction of the beam. In a case such as this the bolts are said to be in a state of single shear (Figure 6-13a) because the shearing force is resisted by one cross section of the bolt. If we had a connection as shown in Figure 6-12, where beams are connecting to either side of a girder, and connected through the girder with bolts, the bolts will be in a state of double shear (Figure 6-13b) since there are shearing forces being resisted across two cross sections of each bolt in such a configuration.

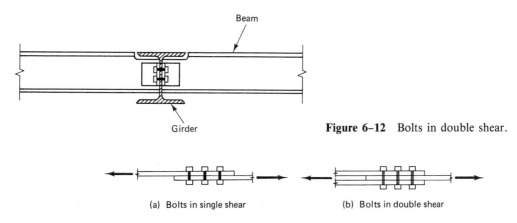

Figure 6-12 Bolts in double shear.

(a) Bolts in single shear   (b) Bolts in double shear

Figure 6-13

In addition to the behavior of fastening systems in beam-to-column and beam-to-girder connections, there are situations where bolts in a connection are subjected to pure tensile forces. For example, in the hanger connection shown in Figure 6–14,

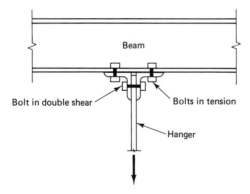

Beam

Bolt in double shear

Bolts in tension

Hanger

Figure 6–14

the bolts connecting the angles to the beam would be in a state of direct tension, while the bolts through the hanger itself would be in a state of double shear. Another type of behavior occurs in a situation like this, which is referred to as *prying action*. In fact, prying action occurs where bolts are in a direct state of tension and the load is being transmitted through the legs of angles or tees. Figure 6–15a shows the possible effect of prying action on a hanger connection, and Figure 6–15b shows the effect of prying action on an angle which may be the tensile side of a moment-resisting connection. The effects of prying action can be quite serious and must be considered in the design of the parts of a connection that transfer tensile forces.

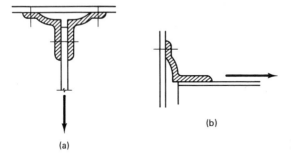

(b)

(a)

Figure 6–15   Prying action.

BOLTS

In steel building structures there are, essentially, two types of fastening systems used. They are bolts and welds. We discuss both of these fastening systems, beginning with steel bolts.

There are three types of bolts that are most commonly used in building structures. They have ASTM designations of A307, A325, and A490. A307 bolts are often referred to a *unfinished, common,* or *machine bolts.* They are the least expensive of the three bolt types mentioned. A307 bolts are used primarily for secondary members such as purlins and bracing, or for small and lightly loaded structures. In larger-scale structures A325 or A490 bolts would be the preferred choice. These bolts are referred to as *high-strength bolts* and have higher tensile and shearing strengths than the A307 type.

When selecting high-strength bolts there are several conditions that must be evaluated. There are two ways that these bolts may be used to connect the parts of a structure. They can be either friction-type connections or bearing-type connections. In the friction-type connection the strength of a joint is achieved by tightly clamping the parts together so that a great deal of friction is created between the parts. The bolts must be tightened to the point where enough friction is created between the parts so that slippage will not occur. Normally, in an architectural structure, friction connections are not required. They are used primarily where structures are dynamically loaded (vibration, impact, etc.), such as bridges and parts of industrial facilities. A bearing connection is based on the idea that the bolt is bearing against the parts that are being connected, and slippage is not an issue as it is in a friction connection. Because slippage is of no concern, a higher allowable shearing stress is assigned to bolts used in a bearing connection.

In addition to the consideration of whether the connection is a friction or a bearing connection, we must also consider whether the threads are included or excluded from the shear plane of the bolt. Obviously, if the threads are within the shear plane, the cross-section resisting shear is smaller than the bolt diameter would indicate. In addition, the threading operation apparently reduces the shearing unit stress that can be resisted. Table 6–1 shows allowable stresses for high-strength bolts. Only bearing-type connections are shown, since these are most common in architectural structures. Based on the values shown it would make little sense in normal situations to have the threads included in the shear plane. However, this condition does occur when thin material is being connected and it is not practical to exclude the threads. Where possible and practical this should be avoided.

TABLE 6–1   ALLOWABLE STRESSES FOR HIGH-STRENGTH BOLTS

Bolt[a]	Tensile stress, F_t (ksi)	Shearing stress, F_v (ksi), bearing-type connection[b]
A325 N	44	21
A325 X	44	30
A490 N	54	28
A490 X	54	40

[a]N, Bearing-type connection with threads *included* in shear plane; X, bearing-type connection with threads *excluded* from shear plane.

[b]The values shown for allowable shearing stress are based on standard round holes (bolt diameter $+\frac{1}{16}$ in.).

It was said, at the beginning of this chapter, that for all practical purposes, rivets are no longer used in building structures. Where this type of fastening is called for, high-strength bolts have totally replaced rivets. There are several reasons for this which also suggest the inherent advantages in bolted connections. Primarily, the advantages are that bolting is quicker, requires less labor, and there is little noise involved as compared to riveting. Bolted connections also provide the opportunity to alter or disassemble a steel frame easily.

WELDING

In the simplest terms, welding is a process whereby two pieces of metal are joined by heating their contact surfaces until they become fluid and allowing them to flow together. Upon cooling, a continuous bond is created between the pieces. Most structural welding today is electric arc welding. In this process, electric current is used which heats an electrode until it becomes fluid and is then used as a filler between the two pieces of metal being connected. In this process the electrode material mixes with the metals being joined in order to develop continuity of the material. When properly designed and executed a welded joint is very efficient and, depending on the state of stress, may be stronger than the base metal.

Electric arc welding was invented in 1881, although not specifically for building structure purposes. In spite of the fact that the steel frame was becoming popular during the first part of the twentieth century, the idea of welding did not make any progress until the 1920s. Even then, when welding was used, there were no changes in the design procedures for structural members. There was no consideration of the continuity created between members. A major breakthrough in this regard was the structural design of the Westinghouse Electric and Manufacturing Company Building in Sharon, Pennsylvania, built in 1926. In this building recognition was given to the fact that welding creates a rigid system, thereby reducing positive moments in members through the introduction of negative moments at the supports.

Much of the delay in the acceptance and development of welding for building structures was because of the unknown qualities and, consequently, fear concerning the reliability of the technique. Also, rigid frame analysis, without the benefit of computers, was quite difficult and tedious. Even though, through the 1930s and 1940s, engineers were coming to accept the procedures of analysis required for all-welded rigid frames, many large cities required special permits for welded structures.

Today, welding and inspection techniques have become quite sophisticated and reliable. In addition, the computer has eliminated much of the burden and the tedium involved in the analysis of a rigid frame structure.

Advantages of Welding

Welding has certain advantages over bolted connections, in two very different ways. An all-welded steel frame may be thought of as a single integral structural unit. In fact, an analogy may be made between the all-welded steel frame and the cast-in-place reinforced concrete frame. Both frames may be designed as completely rigid frames. In a rigid frame structure a savings in the weight of material can be realized because of the continuity between structural members. However, welding, especially field welding, requires extreme skill and sophisticated testing procedures, and is therefore not an inexpensive proccess.

Welding also provides the possibility of visually clean connections. Where a structural frame is to be exposed, connections between the parts may be of great aesthetic importance. With the welding process, many parts of a connection can be eliminated and the welds themselves, where necessary and structurally feasible, may be ground to a smooth finish. In addition, the connections between certain structural shapes, which may be awkward with bolted connections, are quite simple and clean when welding is used. For example, the connection of a wide flange to a rectangular tube column may be accomplished in a variety of ways, but in all cases, welding is the only sensible solution (see Figure 6-16). It may be said that the tube column itself is possible only because of seam welding.

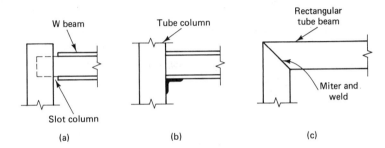

Figure 6-16   Beam to tubular column connections.

Types of Welds

In the broadest sense, there are two types of welds: fillet welds and groove welds. These are shown in Figure 6-17. Fillet welds are the most common in structural connections. Groove welds, although generally stronger than fillet welds, can be somewhat difficult because of the precise fit between pieces that is required. In addition

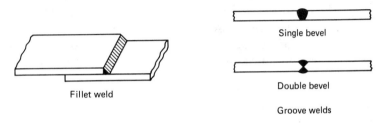

Figure 6-17

to fillet and groove welds, there are slot welds and plug welds, as shown in Figure 6–18. These welds are used in special situations where welding of one plate to another requires some "stitching" between the edges of the plates. Both the groove weld and

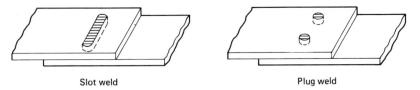

Slot weld   Plug weld

Figure 6–18

plug (or slot) weld may be ground smoothly to be flush with the plates. There are times when, for aesthetic purposes, this may be desirable.

Fillet welds are much stronger in a direct state of tension or compression than they are in their resistance to shearing stresses. Details of fillet welds are shown in Figure 6–19. The common mode of failure of a fillet weld would be in shear, and this would occur along the throat of the weld. Fillet welds generally are made with equal legs, as shown in Figure 6–19. The throat dimension used is based on an angle of 45° at the throat of the weld. The effective area resisting shearing stress is the length of the weld times the dimension across the throat. In the design of welded joints there are, essentially, two welding electrodes that may be used. They are classified as E70 and E80 electrodes. E70 electrodes are used with steels having yield stresses of up to 60 ksi. The ultimate strength of the material deposited by an E70 electrode is 70 ksi.*

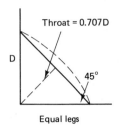

Equal legs

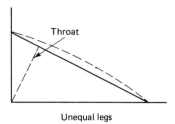

Unequal legs

Figure 6–19   Fillet welds.

In addition to the types of welds discussed, there are a variety of joint types that may be made with welds. There are butt joints, lap joints, T-joints, and others, and some of these are illustrated in Figure 6–20. There are clearly a great variety of possibilities of making welded joints with fillet or groove welds, and the geometrical arrangement is limited only by the designer's judgment and the practicality of the arrangement.

*Sections 1.5.3, 1.14.6 and 1.17 of the AISC Specification gives allowable stresses and provides details of requirements for welds. Those interested in the specific requirements should refer to this source.

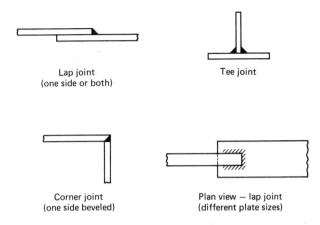

Lap joint
(one side or both)

Tee joint

Corner joint
(one side beveled)

Plan view — lap joint
(different plate sizes)

Figure 6–20

CONCLUSION

There are, virtually, an infinite number of possibilities that could be discussed when considering the various stresses to which parts of connecting items are subjected in a steel structure. The foregoing examples have been presented to, hopefully, create an awareness of the thinking that must be included in the development and design of a suitable connection between the parts of a structure.

The key word in the development of a connection, regardless of how unusual a situation may be, is *judgment*. It must also be said that the importance of a connection between the parts of a structure can hardly be overemphasized. Regardless of the fact that the various members of a structure may be properly designed, it is possible that the connections between the members may prove to be a weak link and, consequently, a hazardous part of the structural system if good judgment and sound practices are not followed.

SUPPLEMENTARY REFERENCES

Detailing for Steel Construction. Chicago: American Institute of Steel Construction, 1983.

Engineering for Steel Construction. Chicago: American Institute of Steel Construction, 1984.

Manual of Steel Construction, 8th ed. Chicago: American Institute of Steel Construction, 1980.

Modern Welded Structures, selections from Vol. I and II; Vols. III and IV, 1965. Cleveland, Ohio: The James F. Lincoln Arc Welding Foundation, 1970 and 1980.

SUGGESTED ASSIGNMENT

Develop a "Joinery Sketchbook."
1. For each of the following, detail at least two ways for simple connections of steel beams to:
   (a) Steel girders (top of beam and girder at the same elevation)
   (b) Steel columns
      (1) Beam framing into column flange
      (2) Beam framing into column web
   (c) Masonry walls
      (1) Beam resting on top of the wall
      (2) Beam framing into the wall
   (d) Poured concrete walls
      (1) Beam resting on top of the wall
      (2) Beam framing into the wall
   For each detail, list the types of stress that should be considered in the design of the connection.
2. Detail a way to provide continuity from beam to beam where the beams frame into a girder. The top of the beams and girder are at the same elevation. Discuss the stresses that should be considered.
3. Detail a simple connection of a beam to a girder where the beam frames into the girder at a 45° angle (in plain view). The top of the beam and the girder are at the same elevation. Discuss the stresses to be considered.
4. Detail at least two ways for connecting a steel column to a concrete footing. Describe these connections in terms of structural behavior and comment on the conditions that would lead to a choice of connection.

7

STEEL FRAMES
AND LATERAL STABILITY

When designing a structure made of individual pieces, such as a steel frame structure, one of the concerns must be the resistance of the frame to lateral forces. The lateral forces that must be resisted are due largely to wind or earthquake. In Chapter 2 maps were shown indicating wind velocities and seismic risk. As one might expect, this information suggests that the intensity of these forces varies geographically. Although the forces due to wind or earthquake are dynamic in nature, we deal with them as equivalent static forces for the sake of convenience. Values of static equivalents are given in building codes and reference should be made to the appropriate code for specific information and design procedures. These values are usually sufficient for most buildings. However, very tall buildings or those with unusual shapes may require more sophisticated procedures, such as wind-tunnel testing and model analysis, to determine the effects of lateral forces.

To be stable, steel frames must be properly braced, not only to prevent failure but also, excessive lateral deflection, which can cause damage to finishes. There are a variety of ways to brace a building to resist lateral forces to which it may be subjected. We will now consider some of these techniques.

BRACING SYSTEMS

When dealing with steel-framed buildings of one or two stories in height, the necessity for supplementary lateral bracing may not be an issue of great importance, although such needs should be evaluated by the designer. For multistory buildings, however, the design may be greatly influenced by lateral forces, especially when the height of the building increases dramatically relative to the least width. Generally, this would become a concern when the building height is in the order of two or more times the least plan dimension of the building. The major concern here would be for lateral stability in the narrow direction of the building.

This is not meant to suggest that steel frames of one or two stories in height should be designed without consideration of the effects of wind or earthquake. Although the amount of lateral deflection in low-rise buildings may not cause catastrophic effects, there may, at best, be damage to finishes and extreme discomfort on the part of the users if the building frame is not sufficiently stiff. Let us now discuss a variety of techniques that may be used, efficiently, to brace buildings of low to moderate heights. We will discuss this from the standpoint of the variety of configurations and techniques that may be used to provide stability. It should be understood that some of these techniques are suitable for resistance to lateral forces whether they are due to wind or earthquake. However, it must be emphasized that although wind and earthquake both produce lateral forces on building frames the response of the building is quite different to each of these phenomena. Wind forces are a function of the exposed surface of the building and are accumulated from the top of the building down to the base. On the other hand, earthquake forces are not a function of the surface area of the building but, rather, a function of the building mass above the foundation level, where such forces are introduced. The most common ways to brace buildings of low to moderate heights is through the use of X-bracing, shear walls, or moment-resisting connections between girders and columns. When considering earthquake forces as the major source of lateral loading, eccentrically placed diagonal bracing or K-bracing may be most suitable. We describe these techniques in the following discussions.

X-Bracing and Diaphragm Infills

The placement and design of X-bracing to resist lateral forces is largely a simple matter of geometry. For example, let us consider the three-bay one-story frame shown in Figure 7–1a, where beams are connected to the columns with simple connections. Because of the inability of simple connections to resist the tendency for the beam to rotate relative to the column, there will be a lateral deformation of the frame due to the application of a lateral force as shown in Figure 7–1b. To resist such deformation, we can include a diagonal member within any one of the bays and tie it back to the base as shown in Figure 7–1c. Considering the direction of the lateral force, this member will be in a state of tension. If, however, the direction of the lateral force is reversed, the member designed as a tension member would go slack and the frame would, again, be unstable. Therefore, another diagonal member would be placed, as shown in Figure 7–1d. The resulting bracing configuration is an ''X.'' Whether the X-brace is placed in one bay, as shown, or two, or all three bays is to some extent arbitrary. The horizontal component of the force in the diagonal must resist the total lateral force applied, assuming that the frame itself will not provide any resistance. If two X-braces are used, the total lateral force to be resisted will

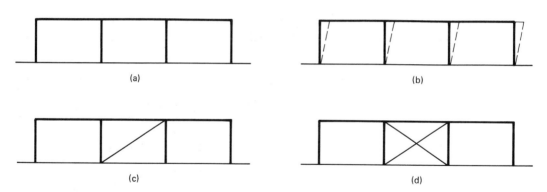

(a)   (b)

(c)   (d)

Figure 7–1

be shared, as it will if three braces are used. Considering seismic forces, however, it is recommended that the bracing pattern be symmetrical within the building, since uniform building stiffness is important when dealing with earthquake forces. Let's now look at a numerical example problem.

Example 7–1

Determine the cross-sectional area required for the diagonal bracing shown in Figure 7–2. Consider that the frame itself cannot resist any of the lateral force. Use A36 steel and the AISC Specification. The braces are to be designed as tension members.

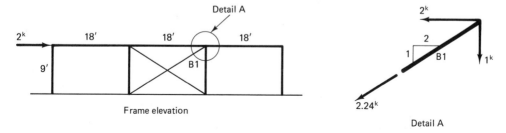

Figure 7–2

Solution   Considering the direction of the lateral force, the brace marked B1 will be in a state of tension and the other brace will go slack. The horizontal component of the tensile force in the brace must equal the applied lateral force, as shown in detail A. Therefore, based on the slope of the brace, the resultant tensile force is 2.24 kips. Considering that the allowable stress, based on the AISC Specification is, $F_a = 0.60\,F_y = 22$ ksi, the required area would be about 0.1 in.^{2}. This may most easily be satisfied by the use of a cable or rod. This problem is now complete.

As suggested by Example 7–1, the size of the bracing required for small buildings will normally be very small. Although the use of very thin cables or rods is very common in small-scale frames, some designers may, for the sake of appearance, wish to choose a member with greater visual substance, especially where the bracing is part of the architectural design.

Referring to detail A of Figure 7–2, it should be noted that there is a vertical component of the force in the brace which adds an axial load to the column in addition to the gravity load. For small buildings this may not be of any great consequence. However, for multistory buildings the additional axial force in the columns may be quite significant since it is cumulative and increases with the building height and number of stories. The same is true for the axial force in the girder due to the horizontal component. This will be demonstrated later.

Considering, again, the three-bay frame of Figure 7–1a, it is also possible to brace the frame against lateral forces with the introduction of vertical walls in the proper locations as shown in Figure 7–3. These are often referred to as *diaphragms*

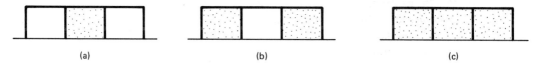

(a)   (b)   (c)

Figure 7–3

or, in tall structures, *shear walls*. Obviously, these diaphragm infills must be made of a material of some structural substance, and they must be positively attached to the frame. With an infill panel of substantial stiffness, the frame will be braced simply because the diaphragm will resist any change to its original rectangular shape. Like X-bracing, the diaphragm infills may be placed in one, two, or all three bays, but again, it is recommended that a symmetrical arrangement be used in deference to the fact that lateral forces may be due to earthquake. It should be clear, from the figures shown, that the arrangement of either diagonal braces or diaphragm infills must be closely coordinated with the architectural requirements and function of the building. Clearly, where openings are required, the use of diagonal braces would normally be precluded. If, however, openings are not terribly large, a diaphragm infill may still be employed if the number of openings and the area is not so large that the integrity of the diaphragm is compromised. Generally, about 30 to 40% of the panel may be open, provided that the arrangement of openings is sensible, and the opening is reinforced, by some means.

We have, thus far, considered frame stability only in the vertical plane. Consideration must also be given to lateral stability of the horizontal plane, especially of a simply connected frame. In Figure 7–4a a steel frame is shown in plan view. Unless all of the frames are braced vertically, there may be some lateral displacement of the unbraced frames. For example, if only the end frames are braced in the vertical plane, there will still be a tendency for the interior frames to move laterally, as shown in Figure 7–4b.

One way to stabilize this situation involves precisely the same principles previously discussed. Considering the lateral forces coming from the direction shown

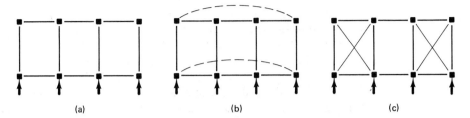

Figure 7-4 Plan view of frame subjected to lateral forces.

in the figure, a tension member may be placed in the horizontal plane as shown. This will tie the interior frame to the end frames, which can be braced in the vertical plane. Considering that the direction of the lateral forces are reversed, the necessary configuration would be an X-brace, as shown in Figure 7-4c. If the roof, or floor, has a concrete deck, which is most common, this may be designed to act as a diaphragm, thereby stabilizing the horizontal plane, provided that the floor system is positively attached to the frame.

To this point we have discussed only steel frames with simple connections. However, connections between beams and columns may be designed as rigid connections, thereby creating a rigid frame. This, in lieu of other forms of bracing, will provide lateral stability by maintaining the original angle between beams and columns. Examples of rigid connections were shown in Chapter 6. We discuss moment resisting connections further a little later in this chapter.

Multistory Frames

The design of structural members in multistory buildings may be more heavily influenced by lateral forces than in one- or two-story buildings. The principles and techniques already discussed for low-profile frames may be applied to multistory frames. X-bracing may be used as shown in Figure 7-5. Again, the placement of X-bracing

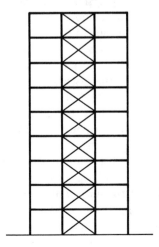

Figure 7-5 X-bracing.

in certain locations may be precluded by the need for openings in the interior for horizontal circulation. In such cases it is possible to use K-bracing as shown in Figure 7–6, as long as the required openings are not too large. Another possibility would be to use knee bracing, as shown in Figure 7–7. Knee braces have the effect, essentially, of maintaining the original angle between girders and columns, thereby providing stiffness to the frame, although not as much as full X-bracing. In lieu of the types of bracing shown, the connections between girders and columns may be designed as rigid or semi-rigid connections. Semi-rigid connections were defined in Chapter 6, and we will discuss these in more detail later.

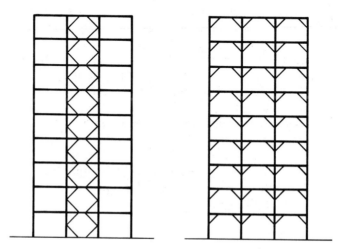

Figure 7–6 K-bracing.   **Figure 7–7**   Knee bracing.

In all cases, and as suggested by the preceding figures, it is important that all floors be braced in a multistory building because there must be a continuous transfer of lateral forces from floor to floor. Should X-bracing, for example, be omitted at one floor, the lateral forces will still find their way through the system (most likely through the floor deck) in order to get to the ground. Although this can be handled through proper design of the floor system as a diaphragm, such techniques should be avoided, especially in earthquake-prone areas.

It was shown previously, in the discussion about diagonal bracing for a one-story frame, that the vertical component of the force in the brace will add axial load to the column. As a building gets taller the magnitude of this additional axial load becomes increasingly important, especially in the design of the columns.

In addition to the tensile force in the brace produced by a lateral force, there will normally be an initial tensile force introduced during the construction process. This is most desirable because it will keep the diagonals from sagging and it will provide the entire building frame with an extra measure of stiffness. It seems that it will now be useful to look at a simple example to show the cumulative effect of lateral forces in a frame with diagonal bracing.

Example 7–2

The six-story frame of Figure 7–8 is subjected to lateral forces, as shown. Determine the force in the braces marked B1, B2, and B3, and the axial force in column C3. Assume that there is no significant initial tensile force in the brace and that the frame does not resist any of the lateral force. The braces are tension members and cannot resist compression.

Solution   The horizontal component of the force in the brace at any level must be equal to the sum of the lateral forces above that level. Therefore, considering the geometry of the brace, shown in detail A:

Roof: $H = 3$ kips, $V = 1.25$ kips, B1 $= 3.25$ kips

Sixth Floor: $H = 9$ kips, $V = 3.75$ kips, B2 $= 9.75$ kips

Fifth Floor: $H = 15$ kips, $V = 6.25$ kips, B3 $= 16.25$ kips

The axial load introduced to column C3 is simply the sum of the vertical components (V) above the fourth-floor level. Therefore, the additional axial force on C3 due to lateral forces is 11.25 kips. If this same process is continued to the lowest column, it will be found that there is an axial force at this level of 45 kips. This represents a substantial accumulation of additional axial load in the column and must be considered in the column design process. The problem is now complete.

More common, perhaps, than the use of X- or K-bracing in tall buildings is the use of shear walls or cores. All tall buildings obviously require space for vertical circulation and mechanical chases. The items that go into these spaces, such as ele-

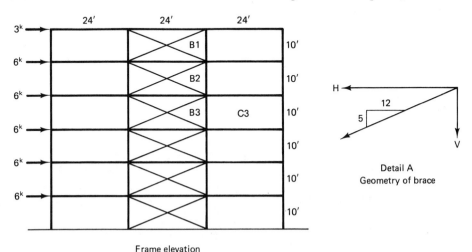

Frame elevation

Figure 7–8

vators, stairs, ductwork, and so on, must, as a general rule, have a great deal of fire protection. Consequently, these items are normally placed within utility cores made of reinforced concrete walls. This forms a large vertical cantilever with great lateral stiffness and vertical load-carrying capacity. Therefore, such an enclosure may be designed to brace the frame. For earthquake forces, the utility core or cores should be placed symmetrically to provide the building with reasonably uniform stiffness. Some suggestions for plan arrangements are shown in Figure 7-9.

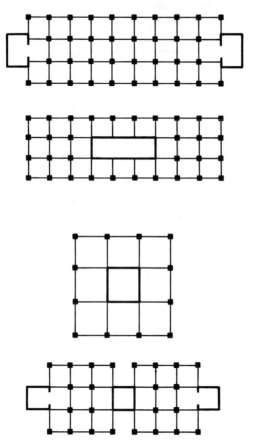

Figure 7-9

In certain building types it may be necessary to provide fire-resistive separation between spaces at fairly regular intervals. These could be reinforced concrete walls which have high fire resistance. Such walls may be used to provide great lateral stiffness to a building frame, in addition to the utility core. These walls are referred to as *shear walls*. It is usually necessary to have openings in shear walls and cores. Openings, if reinforced properly, will not destroy the structural integrity of the shear wall. When using shear walls and/or cores it is best to use them to support the floor framing and carry gravity loads, in addition to lateral loads, as this will increase their stability. Possible arrangements of shear wall systems are shown in Figure 7-10. Again, it is strongly recommended that all such arrangements be symmetrical, especially if the lateral forces to be resisted are due to seismic action.

It should be noted that it is usually not necessary to brace every frame in a building with vertical bracing, as suggested by Figure 7–10b. Normally, the floor system acting as a diaphragm can transfer lateral forces to the braced frames or shear walls. It should also be noted that it is possible to utilize combinations of X-bracing with shear walls or utility cores. Although it has been said many times previously, the key word in any bracing arrangement for multistory buildings is *symmetry*. Buildings that are seriously lacking in reasonably uniform lateral stiffness may suffer from serious torsional effects; that is, rotation in the horizontal plane of the building. This, even though considered in the design process, may cause difficulties when the building is subjected to the energy that may be released by an earthquake.

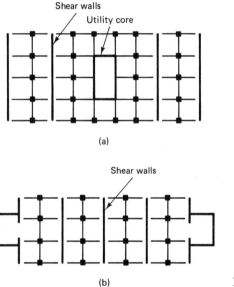

(a)

(b)   **Figure 7–10**

STEEL FRAMES WITH MOMENT-RESISTING CONNECTIONS

It was mentioned earlier in this chapter that frame stability may be achieved through the use of rigid, or moment-resisting, connections between girders and columns. If the connection is designed so that it will resist bending moment, there will be no relative rotation between the horizontal and vertical members, thereby providing resistance to lateral displacement of the frame. This can be achieved through methods of construction previously described in Chapter 6. These are type 1 construction, which is defined as "rigid frame," or type 3 construction, which is defined as "semi-rigid framing." We will discuss both of these types of framing in a bit more detail later. In either case it is important to determine the magnitude of the moments, due to lateral forces, that must be resisted.

We will study an approximate method to determine the moments due to lateral forces, which is known as the *portal method*. The portal method, although an approximate method, is reasonably sufficient for frames of regular proportions and up to 20 or 25 stories in height.

There are, indeed, accurate methods of analysis available that can be done by computers. However, computer programs generally require some initial input in regard to the stiffness of columns and girders. Consequently, although the portal method is an approximation, it does serve a useful purpose for making preliminary determinations for computer input.

The Portal Method

To make an approximate analysis using the portal method, the general step-by-step procedure is as follows:

1. Determine the wind pressure acting on the surface of the building. This would be the result of the recommended wind pressure suggested by the appropriate building code.

2. Collect the uniformly distributed wind pressure and show it as a concentrated load at the floor level.

3. Show each bay as a separate portal with the total lateral load equally divided among them, assuming that the bay widths are equal. If the bay, or portal, widths are not equal, one approach is to assume that the total lateral load, at a floor level, is distributed to each portal in direct proportion to the floor area being supported. In this procedure each interior column will be shown with each of the portals it shares.

The following assumptions are used in the procedure:

1. The point of contraflexure of each column is at the midheight of the story.

2. The point of contraflexure of each girder is at the midspan.

3. The shears on the internal columns are equal, and the shear on each external column is equal to one-half of the shear on an internal column. In other words, the lateral force at any level divides in the ratio of one part to each external column and two parts to each of the internal columns. For the purpose of calculating this distribution, use the following:

$$\text{number of columns } (n) = 2n - 2$$

4. Using the assumptions above, proceed to analyze each portal on each floor and calculate the shears and moments for the columns and girders.

The following must also be considered in the procedure:

1. The lateral load is cumulative from the top of the frame to the base of the frame.

2. The lateral load is considered to accumulate directly below the floor in question.

3. The maximum girder moment produced by the lateral force occurs where the girder frames into the column. Because the lateral load is cumulative, so will be the girder moment.

Example 7-3

To demonstrate the procedures outlined, let us use the six-story frame shown in Example 7-2 and analyze the conditions at the roof level and the sixth-floor level. Let us consider that this is an interior frame and that the frames are spaced at 20 ft apart. Since the method is an approximate method, it will be more than sufficient to round off numbers to, at most, one decimal place.

Solution   Following the procedures, Figure 7-11a shows an elevation of the frame with the uniformly distributed wind pressures on the surface recommended by the appropriate building code. As one would expect, these equivalent static

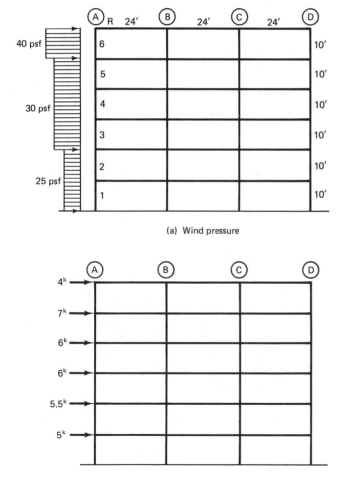

(a) Wind pressure

(b) Wind pressure as concentrated loads

Figure 7-11   Interior frames spaced at 20 ft.

pressures vary with the height of the frame. We then collect these uniform pressures and show them as concentrated loads at the floor levels, as shown in Figure 7-11b. Since this is an interior frame and the frames are spaced at 20 ft apart, the value of the concentrated load at the roof level is

$$P = (5 \text{ ft})(20 \text{ ft})(40 \text{ lb/ft}^2) = 4000 \text{ lb} = 4 \text{ kips}$$

At the sixth-floor level,

$$P = (5 \text{ ft})(20 \text{ ft})(40 \text{ lb/ft}^2) + (5 \text{ ft})(20 \text{ ft})(30 \text{ lb/ft}^2) = 7000 \text{ lb}$$

This same procedure is followed down to the second-floor level.

We then proceed to show each bay as a separate portal, each carrying an equal share of the lateral force at each floor level, as shown in Figure 7-12.

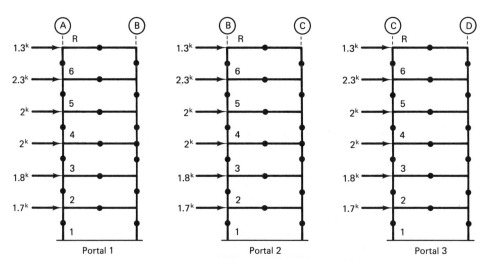

Figure 7-12

Then, with each portal, we take a series of free-body diagrams through the points of contraflexure, showing the concentrated load acting at the floor level. This is shown in Figure 7-13. The applied lateral load in each portal is assumed to be resisted as a shearing force equally divided between the two columns. It should be noted, in the breakdown of the individual portals, that column B and column C are shown twice. This may be thought of as one-half of each column functioning with the adjacent portal. This procedure is followed for all three portals. The shearing forces in the columns satisfy the conditions for static equilibrium in the horizontal direction ($\Sigma F_x = 0$). However, in looking at these free bodies it will be seen that rotation will be taking place. To counteract this rotation there will be vertical forces involved at the point where the free body was cut. To satisfy $\Sigma M = 0$, take moments at one of the points of contraflexure to determine the vertical force. Using portal 1, for example, the computation would be

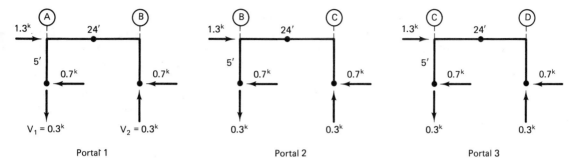

Figure 7-13 Roof-level free bodies.

$\Sigma\,M_x = 0 = (1.3 \text{ kips})(5 \text{ ft}) - V_1(24 \text{ ft})$ and $V_1 = 0.3$ kip downward
Therefore, $V_2 = 0.3$ kip upward ($\Sigma\,F_y = 0$).

Note that using the portal method assumptions, when the portals are put back together, the vertical force shown in column B for portal 1 is negated by the vertical force in column B for portal 2. The same is true for column C. In the portal method, the net axial force on the interior columns, due to lateral forces, is zero and the axial force in the column on the leeward side is compressive and on the windward side, tensile. Of course, it should be realized that the lateral forces may change direction. Therefore, when determining axial forces we would be concerned with the additional compressive force on the exterior column.

Let us now go to the sixth floor and make a similar analysis. Again, a free body, including the level being investigated, is taken through the assumed points of contraflexure as shown in Figure 7-14. The shearing forces from the

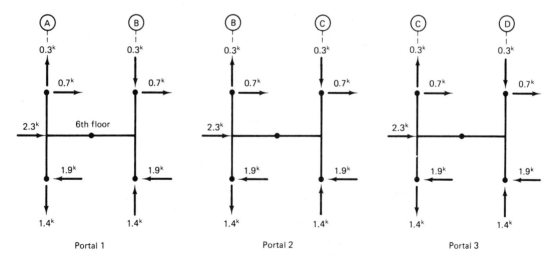

Figure 7-14 Sixth-floor free bodies.

portals above, which was shown as a *reaction,* is now shown as an *action* on the columns. In addition, another lateral load is introduced at the sixth-floor level. Therefore, the total shear on any of the portals is equal to the summation of forces from above, and the lateral force being introduced at the level in question. The total lateral force is assumed to divide equally as reactions in the columns below the floor level. This is shown for all portals. In addition to the lateral forces coming from above, we must also remember to include the vertical forces. These vertical *reactions* at the roof-level portals must be shown as *actions* on the sixth-floor portals. The necessary vertical reactions for static equilibrium are shown in the figure. The student can now proceed with these techniques down to the base of the building. It should be emphasized that the lateral loads are cumulative and the total shear at any level is equal to the sum of the lateral forces above that level.

Let us now proceed to determine the column and girder moments at the roof and sixth floor. The moment in the exterior column below the roof level is simply the product of the lateral shear and the lever arm to the top of the column. Therefore,

$$M = (0.7 \text{ kip})(5 \text{ ft}) = 3.5 \text{ ft-kips}$$

The moment on the interior column, directly below the roof level, is equal to twice the moment on the exterior column, since the shear on the interior column is twice that on the exterior. At the junction between the girder and the exterior column, the moment in the girder must also be 3.5 ft-kips. The girder moment at the interior column, however, does not double because, while there is only one column at this point, there are two girders that will equally share the moment. It may help to clarify the matter to look at a moment diagram superimposed on the frame shape. We will use the free bodies of the portals at the roof level, and this is shown in Figure 7–15.

Going to the sixth-floor level, shown in Figure 7–14, the moment in the exterior column just above the floor is, again, 3.5 ft-kips. Just below the floor the column moment is 9.5 ft-kips. The girder moment is the sum of the column moments, which is 13 ft-kips. The moment on the interior columns is twice that of the exterior columns. The connections between girders and columns would have to be designed to resist, at minimum, the value of the girder moment. For our purposes this example problem is now complete.

Let us now briefly discuss a few points about type 1, type 2, and type 3 construction, as defined by the AISC Specification. Type 1 construction is rigid frame construction, where the girder-to-column connections are completely rigid. Essentially this means that the connection will have a moment capacity equal to that of the girder. Type 2 construction means that the girder (or beam)-to-column connections are flexible and will not resist rotation under gravity loads. According to the AISC Specification, connections to resist wind moments may also be provided with type 2 construction, provided that the girders can carry the full gravity load as simple spans. Type 3 construction, which is semi-rigid construction, means that the connec-

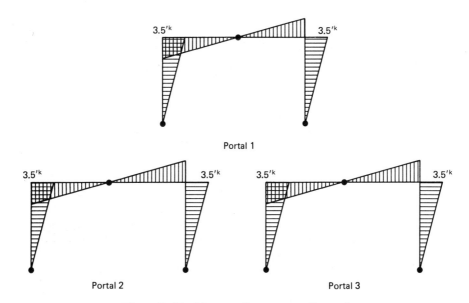

Figure 7-15 Moment diagram—roof portals.

tions will have a known moment capacity which is less than required for full end restraint. It is somewhere between type 1 and type 3 framing. Since the "known moment capacity" of type 3 framing may be the moment due to wind, the difference between type 2 framing with wind moment connections and type 3 framing may be largely philosophical.

In a frame, such as the one of the preceding example, it is most likely that type 2 framing with wind connections would be most suitable. The connections between girders and columns would be designed to resist the moments produced by the lateral forces. Philosophically, we may think of such a situation as one where the girder-to-column connection is a simple connection under gravity loads and the moment-carrying part of the connection is inactive when lateral forces are not acting. When lateral forces are acting the simple connection continues to do its job and the wind connection goes into action. Actually, since both parts of the connection are in place at all times, both will be acting to some degree even under the action of only gravity loads. Typical connections of the type being discussed were shown in Chapter 6.

When designing the girder at the upper stories it will, no doubt, be the moment due to gravity loads that will dictate the member size. However, as we go down to the lower levels the moment due to wind may be the controlling design moment. In fact, it may be necessary to superimpose the moment diagrams for lateral forces and gravity loads in order to determine the largest design moment. Normally, it will suffice to take the larger of the two as the design moment. It should, however, be recognized that somewhere between the maximum moment at the end of the girder due to wind and the maximum simple span moment at the midspan there will be a combination of the two that is slightly larger than either.

Fortunately, as mentioned in Chapter 3, the AISC Specification allows an increase in allowable stresses of 33% when these stresses are produced by wind or seismic loading acting alone or in combination with the gravity loads. This is allowed provided that the section selected on this basis is not less than that required for the gravity loads alone.

The Cantilever Method

It was shown in the portal method that based on the assumptions, the direct axial load due to wind on the interior columns is zero. While the portal method does yield a value for the direct load on the exterior columns, there is another approximate method, known as the *cantilever method,* that yields somewhat more accurate results for the column axial loads. Also, it provides a value for the axial load on the interior column which the portal method does not. We will briefly look at the cantilever method only insofar as it provides us with direct axial loads on the columns. In the cantilever method the determination of the axial force in a column is based on the assumption that the forces will vary in proportion to the distance from the center of gravity of the column group. To determine the direct load on the columns, a free body of the frame is taken through the point of contraflexure in the columns which, like the portal method, is assumed to be at the midheight of the story. This operation can be performed at any level of the frame.

Example 7-4

Use the frame and the loads of Example 7-3 and determine the direct load on the columns at the lowest level.

Solution   The frame and the loads are shown in Figure 7-16a. The direct axial loads, considering the wind direction as shown, will be compressive on the leeward side and tensile on the windward side. Assuming that the magnitude of the axial forces will vary in proportion to the distance from the center of the frame, axial forces in terms of P are shown in the figure. Taking a free body of the upper part of the frame through plane A–A and taking moments, we have

$$\Sigma M_x = 0 = (4 \text{ kips})(55 \text{ ft}) + (7 \text{ kips})(45 \text{ ft}) + (6 \text{ kips})(35 \text{ ft}) + (6 \text{ kips})(25 \text{ ft})$$
$$+ (5.5 \text{ kips})(15 \text{ ft}) + (5 \text{ kips})(5 \text{ ft}) + (P)(24 \text{ ft}) - P(48 \text{ ft}) - 3P(72 \text{ ft})$$
$$0 = 1002.5 \text{ ft-kips} - 240P \qquad P = 4.2 \text{ kips}$$

Therefore, the compressive force due to lateral loading, in the exterior column on the leeward side is $3P = 12.6$ kips.

For the sake of a bit more practice, let us now determine the axial force in the columns, due to wind, between the fifth and sixth floors. The free-body diagram is shown in Figure 7-16b. Therefore,

$$\Sigma M_x =$$
$$0 = (4 \text{ kips})(15 \text{ ft}) + (7 \text{ kips})(5 \text{ ft}) + (P)(24 \text{ ft}) - (P)(48 \text{ ft}) - 3P(72 \text{ ft})$$
$$0 = 95 - 240P \qquad P = 0.4 \text{ kips} \qquad 3P = 1.2 \text{ kips}$$

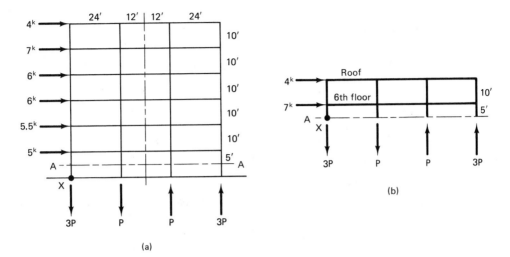

Figure 7–16

Comparing this to the value determined by the portal method, this value is somewhat less, as is the value determined at the lower level. The values determined by the cantilever method, however, are more accurate and the method also gives us values for the interior columns. This example is now complete.

When designing a rigid frame, or one of type 2 construction with wind connections, one of the issues that must be considered in the design of the columns is the fact that without a positive system of diagonal bracing, the frame will have the tendency to deflect laterally. This is referred to as *sidesway*. In such cases the factor that we use to determine the effective length of the column (K), as defined in Chapter 5, may not be easy to determine. In fact, the effective length factor (K) could be significantly more than 1.0. The commentary on the AISC Specification (Section 1.8) provides a method for determining the unbraced length factor to be used when frame sidesway is uninhibited. Once this has been determined, the design procedure for columns subjected to axial load and bending moment presented in Chapter 5 may be followed.

TUBE STRUCTURES

The braced frame and shear wall structures that have been discussed to this point are generally efficient for buildings of about 30 to 40 stories in height. For tall buildings, however, a more recent innovation in framing may be most efficient in the design of taller buildings, where the effects of lateral forces become increasingly intense. These are known as *tube structures*. We discuss these systems next, albeit briefly.

Essentially, a tube structure is one where the exterior parts of the exterior frame of the building are rigidly connected, thereby forming a very rigid "cage," or peripheral tube. In a way, we have already touched on the tube structure concept in our previous discussion about an interior core that resists lateral forces. Such an interior core is analogous to the tube structure we are now discussing. In fact, in some tube structures the exterior tube is connected to an interior tube and the two designed to behave as an integral structural system.

There are several varieties of tube structures. These can be categorized into four major groups; the framed tube, the trussed tube, the tube-in-tube, and the bundled tube. There are possible variations within each group.

Framed tube.   The framed tube is one where the entire exterior frame is made of closely spaced columns rigidly connected to spandrel beams. In essence, because of closeness of the spacing, the framed tube may be thought of as a bearing wall. It is very rigid and quite efficient in resisting shear and bending produced by lateral forces. The structure within the exterior frame is designed to take only gravity loads. A prime example of a framed tube is the 110-story World Trade Center in New York.

Figure 7-17   Standard Oil Building, Chicago.

A variation of the framed tube is known as a perforated wall tube, or shell tube. In this system, again, the exterior parts are rigidly connected to form an extremely rigid cage. The exterior is made of deep spandrel girders and wide columns, made of large steel plates. The amount of openings in the surface is relatively small, and represents about 30 to 40% of the surface area. During construction, and before fenestration, the large steel plates on the exterior are predominate and the walls almost appear to be of solid steel. An example of this sort of tube structure is the Standard Oil Building in Chicago, shown in Figure 7–17.

Trussed tube.   In a trussed tube structure, such as the John Hancock Center in Chicago shown in Figure 7–18, the exterior cage combines vertical, horizontal, and diagonal members, all rigidly connected. The diagonals carry gravity loads as well as lateral loads. They may be thought of as major columns that are diagonally oriented. This technique imparts a great deal of stiffness to the system. The trussed tube has proven to be very efficient. For example, the structural steel weight in the Hancock Center averages out to about 30 lb/ft^2. For a building of similar propor-

Figure 7–18 John Hancock Center, Chicago.

tions of rigid frame construction with supplementary bracing the steel weight would probably be somewhere around 60 lb/ft^2.

Tube-in-tube. In both of the tubular systems discussed thus far, the design is based on the concept that only the exterior frame is resisting the shear and bending moments produced by lateral forces. In fact, special care is taken to avoid the transfer of any lateral force to the interior parts of the building. In a tube-in-tube structure, however, the exterior tube and an interior tube are designed to act together. Because interior tubes are generally very slender, they are designed to carry the shear produced by the lateral force, and the exterior tube, because of its large width, is designed to resist the bending moment.

The tube-in-tube concept may be most efficient in a building where the amount of openings in the exterior frame is rather large, thereby diminishing its ability to resist shear. Lateral shearing forces may be transferred to the interior tube through a stiff floor system. The interior tube may simply be the concrete shaft which houses stairs, elevators, and so on.

Figure 7-19 Sears Tower, Chicago.

Bundled tube.   The bundled tube concept was used in the 110-story Sears Tower in Chicago, shown in Figure 7–19. The system basically consists of nine framed tubes with shear walls between exterior walls. The arrangement is shown, schematically, in Figure 7–20. The bundled tube concept allows each module in the building to be designed as an independent framed tube. In the Sears Tower the tubes are each 75 ft × 75 ft arranged in a square at the base and they rise to various heights. Only two of the tubes reach the full height.

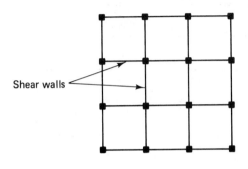

Shear walls

Figure 7–20   Plan view of bundled tube. Each square represents an individual tube. In the Sears Tower, nine tubes are bundled together at the lower level.

SUPPLEMENTARY REFERENCES

DEGENKOLB, HENRY J. *Earthquake Forces on Tall Structures,* Booklet 2717A. Bethlehem, Pa.: Bethlehem Steel Corporation, 1977.

"Drift in High-Rise Steel Framing," *Progressive Architecture,* April 1972, p. 98.

GREEN, NORMAN B. *Earthquake Resistant Building Design and Construction.* New York: Van Nostrand Rembold, 1981

LIBBY, JAMES R. "Eccentrically Braced Frame Construction—A Case History," *AISC Engineering Journal,* Vol. 18, No. 4, Fourth Quarter 1981, pp. 149–153.

"Optimizing Structural Design in Very Tall Buildings," *Architectural Record,* August 1970, pp. 133–136.

POPOV, EGOR P. "An Update on Eccentric Seismic Bracing," *AISC Engineering Journal,* Vol. 17, No. 2, Third Quarter 1980, pp. 70–71.

POPOV, EGOR P. "Seismic Steel Framing Systems for Tall Buildings," *AISC Engineering Journal,* Vol. 19, No. 3, Third Quarter 1982, pp. 141–149.

"Tall Buildings—Type Study," *AIA Journal,* January 1973, pp. 15–40.

WIEGEL, ROBERT L., Editor. *Earthquake Engineering.* Englewood Cliffs, N.J.: Prentice-Hall, Inc., 1970.

WILSON, FORREST, and BURNS, JAMES T. *Seismic Sculpture* Progressive Architecture. December 1968, pp. 84–91.

PROBLEMS

(7.1.) The frame shown is subjected to lateral forces. The X-bracing can resist only tensile forces. Based on A36 steel and the AISC Specification, determine the cross-sectional area required for the brace at the lower level. Assume that L/r requirements will be satisfied.

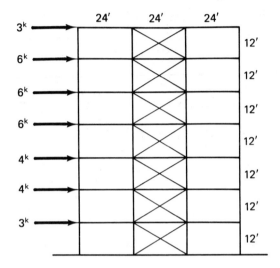

Framing elevation

(7.2.) **(1)** The braces shown can resist compression as well as tension. Determine the compressive force in the braces at each level.

(2) Determine the wide-flange section required for the brace at the lower level, based on A36 steel and the AISC Specification.

(3) Determine the axial load in column C1 due to the lateral forces.

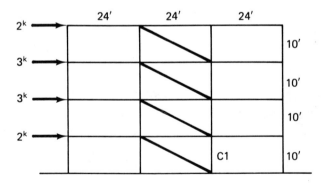

Framing elevation

(7.3.) The X-bracing shown can resist only tension. The braces must resist all of the lateral force. Determine the cross-sectional area required for the braces at the lower level, based on A36 steel and the AISC Specification.

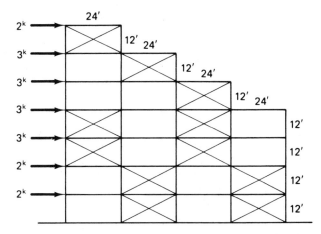

Framing elevation

(7.4.) **(1)** The X-bracing shown resists all of the lateral force, and can only resist tension. Determine the tensile force in each brace.

(2) Determine the axial load in column C1 due to the lateral forces.

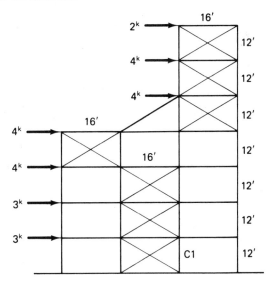

Framing elevation

(7.5.) *Frame Analysis.* The frame shown is to be designed based on type 2 construction with wind connections to resist the lateral forces.

(a) Using the portal method, determine the girder and column moments at each level.

(b) Determine the direct axial load, due to the wind forces, in the first-floor column on the leeward side.

(c) Compare the result of part (b) to the direct axial load found using the cantilever method.

(d) Considering the framing plan shown, for a typical interior bay, and that the total floor load is 200 lb/ft², design a typical girder at the fifth-floor level.

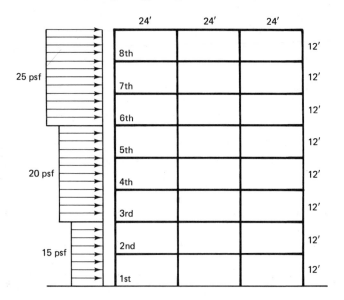

Framing elevation

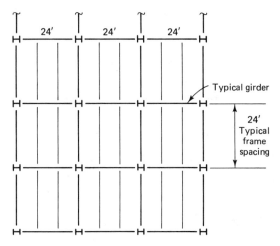

Partial framing plan

8

PLASTIC DESIGN IN STRUCTURAL STEEL

All previous work in this book has been based on the allowable stress design method. This method was first discussed in Chapter 2. Also, in Chapter 2, a newer method of design for structural steel members known as the plastic design method was described. In this chapter we deal with the plastic design process.

Although plastic design in steel is not the most commonly used design method for structural steel buildings, it does provide some advantages over designs made by the allowable stress method.* In general, plastic design greatly simplifies the computations involved in the design of statically indeterminate structures. An additional advantage is that the plastic design process may provide for substantial savings in steel tonnage. On the other hand, however, as we progress through this chapter it should become clear that the plastic design method has no meaningful application when dealing with statically determinate beams (i.e., simply supported). Although the plastic design method may be used for rigid steel frames, we limit our discussions to the analysis and design of only beams with restrained ends. The central purpose of this chapter is twofold. On the one hand it is intended to provide the reader with the fundamental understanding of the plastic design method and the vocabulary involved in the procedures. On the other hand, studying the theory, even though in a rudimentary way, should provide the reader with a sense of the unique properties of steel and its reserve strength.

*Actually, the use of this method can be considered rare in the United States. There are few indications that it will become common in the future.

THE BASIS FOR PLASTIC DESIGN

The plastic design method takes advantage of the *ductility* of structural steel. In particular, *mild* structural steel has some unique properties that form the basis for the philosophy which underlies the method. To illustrate this, we will use the idealized stress versus strain diagram of Figure 8-1. This is shown for grade ASTM A36 steel, which will be our sole consideration. As shown in the figure, when steel reaches its yield stress (F_y), strains will continue to increase without any significant increase in stress. Stresses up to the yield point are said to be within the elastic range of the material and the steel will behave elastically. The range beyond this, as shown in the figure, is known as the *plastic range*. The idea that the stress remains constant throughout this range of strains provide the basis for the development of plastic theory. At a strain of about 10 to 15 times the initial yield strain, the steel specimen in the testing machine will "neck-down" significantly and, again, it will exhibit the ability to resist stress. This range is known as the strain hardening range, and we will not be concerned with behavior in this range. It is interesting, however, to note that the specimen will break at a strain of about 250 to 300 times the initial yield strain. This represents an extraordinary amount of deformation and emphatically shows the great ductility and reserve strength of steel. It is difficult to break a piece of steel except under laboratory conditions. Fracture of a structural steel member in a building is unheard of as a mode of failure.

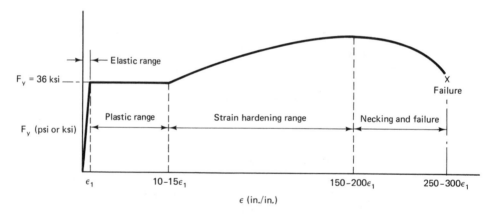

Figure 8-1 Generalized stress versus strain diagram—A36 steel.

Let us now discuss how the property of constant stress within the plastic range allows us to take advantage of the great ductility of steel. To this end we consider a uniformly loaded beam with fixed ends, as shown in Figure 8-2. There will be negative bending moments at the fixed ends and a positive moment at the midspan. If the load on the member is such that the section is stressed within the elastic range, the values of negative and positive moments will be as shown on the moment diagram.

Let us suppose that we are using ASTM A36 steel. Referring back to Chapter 4, the allowable stress that we would design for is 24 ksi, which is clearly below the

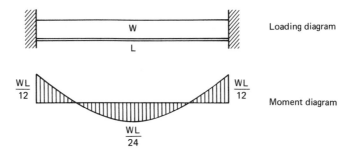

Figure 8-2 Uniformly loaded beam with fixed ends.

yield stress. Now imagine that the load is increased, with a consequent proportional increase in the bending moment. Let us consider that the extent of the increase is such that it will produce yielding at the outermost fiber of the cross section at the points of maximum moment. In this case this will occur at the points of negative moment (the supports). The stress variation diagram for this condition is shown in Figure 8-3. A corresponding strain variation diagram is also shown indicating that the strains at the outermost fibers of the cross section are ε_1. Referring to the stress and strain diagrams of Figure 8-3, it should be realized that stresses between the neutral axis and the outermost fibers of the cross section are still within the elastic range, and consequently, the strains will be less than ε_1.

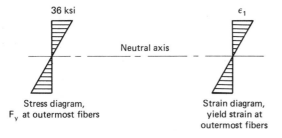

Stress diagram,
F_y at outermost fibers

Strain diagram,
yield strain at
outermost fibers   **Figure 8-3**

Now suppose that the load is further increased. The strains will continue to increase proportionately as the moment increases, but there will be no stress increase beyond the yield stress, within the plastic range.* Consequently, the fibers where strains are greater than ε_1 will stress up to the yield stress, and this is indicated in the stress diagram of Figure 8-4. Now consider that the load is again increased and that the strains will continue to increase proportionately with the distance from the neutral axis. Consider further that the strains have become so large that a strain level of ε_1, as shown in Figure 8-5, occurs, for all practical purposes, at the neutral axis.†

*The fundamental principle that initially plane sections remain plane during bending with a consequent proportional variation in *strain* from the neutral axis to the outermost fibers is true even beyond the elastic range.

†Obviously, this cannot occur exactly at the neutral axis, but it is close enough to be considered in this manner.

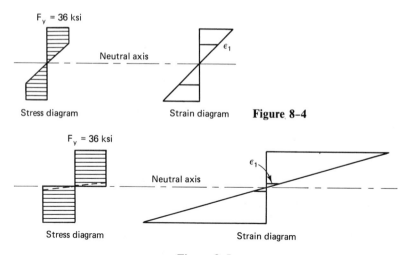

Figure 8-4

Figure 8-5

The corresponding stress variation will be as shown in the figure, where all of the fibers throughout the cross section have yielded. At this point it may be said that the cross section, where this situation occurs, has been totally utilized and cannot provide any more stress to resist any further increase in resisting moment capacity. At this stage the section will have rotated significantly and an analogy can be made to a hinged condition. In fact, this condition is referred to as a *plastic hinge*. When a plastic hinge forms, the section cannot provide any increase in resisting moment. The moment that occurs at the plastic hinge is stored as a constant. In a real hinge, of course, no moment could ever be resisted and the constant value is zero. With reference to Figure 8-5, it may be said that a plastic hinge forms when all fibers have yielded and the "reservoir" of reserve strength of the cross section has been filled. The maximum possible value of the resultants of compression and tension have been achieved, which represents the maximum possible resisting moment capacity available in the section.

All of the action described occurred at the point where we started with maximum values. These were the negative moments at the fixed ends. As we were increasing the load on the beam, the positive moment was also increasing in proportion to the increase in the load. But the positive moment started out, in this case, at one-half the value of the negative moment. Consequently, when plastic hinges form at the end sections the stress diagram for the positive section will look like that shown in Figure 8-3. In fact, the positive moment section is still behaving elastically. Therefore, the deformed shape of the beam will look like that shown in Figure 8-6. Note that after the plastic hinges have formed at the end sections the deformed shape is very much like that of a simply supported beam, still maintaining elastic curvature.

Figure 8-6 Deformed shape.

While the end sections have fully yielded and cannot provide any increase in resisting moment capacity, the deformed shape of the member suggests that more load is required for failure to occur. If we now keep increasing the load, the moments at the ends will remain constant at their maximum possible values, but the midspan section will be able to provide further resistance. It will do this until all fibers at the maximum positive moment section have yielded. At this stage a third plastic hinge will form at the midspan and the deformed shape will be as shown in Figure 8–7. This configuration is referred to as a *three-hinged collapse mechanism* and signifies the limit of usefulness of the beam. Three hinges in a member spanning horizontally between two supports is an unstable configuration and the beam can no longer resist any further increase in load. The values of the moments that are stored at the hinges are equal to each other, since the stress variation (and, consequently, the resisting moment capacity) at each of the three sections is identical. The moments at these sections are referred to as the *plastic moments (M_p)*.

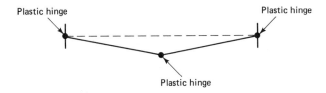

Plastic hinge

Plastic hinge

Plastic hinge

Figure 8–7   Three-hinged collapse mechanism.

Let us summarize, briefly, what has happened in the process just described. We started with a member loaded only to the degree that the sections of maximum negative and positive moments were stressed within the elastic range of the material. The load was then increased to the point where all fibers yielded at the end section and plastic hinges were formed. At locations away from the end sections the beam was still behaving elastically and was therefore capable of carrying more load. The load was then increased in order to form the third plastic hinge, which signifies that the *limit of usefulness* has been reached.

General Expressions for Maximum Moments

Considering the conditions described previously, it should be clear that when the limit of usefulness is reached, the negative moments must be equal to each other and, in turn, equal to the maximum positive moment. Therefore, the general expression for the value of the plastic moment (M_p) for the uniformly loaded beam with restrained ends is equal to $WL/16$. This is so because the total value of the negative and positive moments can never exceed the simple beam bending moment value, which, for a uniformly loaded beam, is $WL/8$.

General expressions for the value of M_p for members with restrained ends are shown on Data Sheet A3. For members with both ends restrained the value of the plastic moment will always be one-half of the simple span moment regardless of the loading pattern. Also given are general expressions for conditions where the member is restrained only at one end. The other end is a real hinge. This condition occurs, for example, in the end span of a continuous beam.

Let us now determine how, for any loading pattern, we would determine the plastic moment value for a beam where one end is restrained and the other end is a hinge. It should be noted that in every case the limit of usefulness for a beam is signified by the occurrence of three hinges. When a member is restrained at both ends, all three hinges are plastic hinges with equal values of plastic moment (M_p). Where one end is a real hinge, this means that only two plastic hinges must develop for the collapse mechanism to occur. We consider next several cases shown on Data Sheet A3 and derive the general expressions shown for beams with only one end restrained. We will, later in this chapter, apply these expressions and the fundamental concepts of plastic design to the design of continuous beams.

Let us consider the case of a beam with one end restrained and loaded as shown in Figure 8-8. The reactions are given in the figure.* To determine the general expression for the plastic moment we will use the basic ideas that the difference in bending moment between any two points is equal to the area under the shear diagram between the same two points, and that the moment is maximum where the shear diagram crosses the baseline. With this in mind and referring to Figure 8-8, it should

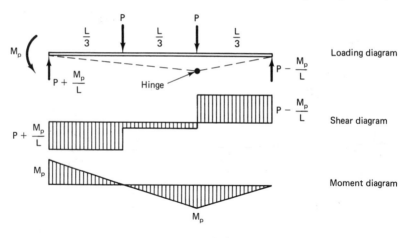

Figure 8-8

be clear that one plastic hinge will occur at the restrained end and we will have a negative M_p at that location. At the unrestrained end we have a real hinge with a bending moment value of zero. The location of the third plastic hinge with a positive M_p is relatively simple to determine in this case. The shear diagram clearly crosses the baseline under the load closest to the unrestrained end, and this is where the posi-

---

*Although it was shown in Chapter 4, it may be useful to point out again, that the reaction for a beam with a negative moment applied at the end is equal to the simple span reaction minus the value of M/L at the unrestrained end and plus M/L at the end where the moment is applied.

tive M_p occurs. To evaluate the plastic moment, the area under the shear diagram is computed as follows:

$$M_p = \left(P - \frac{M_p}{L}\right)\frac{L}{3} = \frac{PL}{3} - \frac{M_p}{3}$$

and

$$M_p = \frac{PL}{4}$$ (as shown on the data sheet in the Appendix)

   In the preceding case, the location of the plastic hinges and the location of the positive M_p were relatively simple to determine. In all cases the location of the negative M_p will always be at the support. In some loading situations, however, the location of the positive M_p may not be so obvious when dealing with beams restrained only at one end. To demonstrate the manner in which such a condition would be handled, let us consider the beam restrained at one end and loaded with equal loads placed at the quarter points, as shown in Figure 8–9. The end reactions for this condition are shown in the figure. There will be a plastic hinge at the restrained end and this is the point of the negative M_p . A second hinge is a real hinge at the left-hand

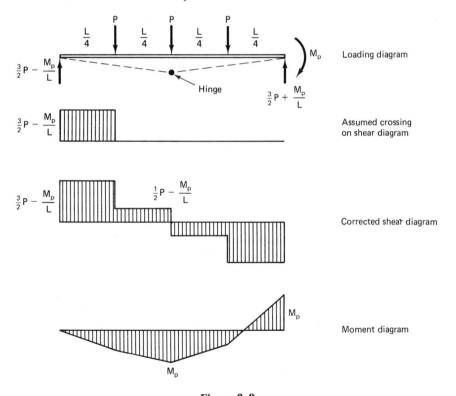

Figure 8–9

end. Considering that the third plastic hinge will form at the point of maximum positive M_p, this will occur where the shear diagram crosses the baseline. In the case being considered this will clearly occur at one of the point loads. However, it is not obvious exactly under which load this will occur. To deal with this, we must make an assumption regarding the point where the shear diagram crosses the baseline. For this case let us assume that the crossing will take place under the first point load from the left-hand end. We will now proceed to evaluate M_p by computing the area under the shear diagram.

$$M_p = \left(\frac{3P}{2} - \frac{M_p}{L}\right)\frac{L}{4} = \frac{3PL}{8} - \frac{M_p}{4}$$

$$\frac{5M_p}{4} = \frac{3PL}{8} \quad \text{and} \quad M_p = \frac{3PL}{10} = 0.3PL$$

Given the value of M_p, based on the assumption, this shows that the shear diagram does not cross at the assumed location and that the computed value is incorrect. It should be obvious now that the shear diagram will cross the baseline under the point load in the middle of the span. The corrected shear diagram is shown in Figure 8–9, along with the moment diagram. The computations for the correct value for M_p, based on the area under the shear diagram, are

$$M_p = \left(\frac{3P}{2} - \frac{M_p}{L}\right)\frac{L}{4} + \left(\frac{1}{2}P - \frac{M_p}{L}\right)\frac{L}{4}$$

$$M_p = \frac{PL}{2} - \frac{M_p}{2} \quad \text{and} \quad M_p = \frac{PL}{3}$$

This value is given on Data Sheet A3. Also shown on the data sheet is the case of a uniformly loaded beam with only one end restrained. The value for M_p is given. Let it be a challenge to the student to derive that value and to determine the location of the positive M_p.

GENERAL EXPRESSIONS FOR ANALYSIS AND DESIGN

As shown earlier in this chapter, plastic design is based on the idea that all fibers within a cross section have yielded and that the yield stress is a constant value within the plastic range. Based on the uniform unit stress distribution, we proceed now to develop necessary information relating to a cross section subjected to a uniform unit stress in tension and compression. This will provide us with general expressions to be used in the design process.

Location of the Neutral Axis

Referring to Figure 8–10, which is a stress diagram for a section stressed to the yield point throughout, we know from the requirements for static equilibrium that the two forces which constitute the internal couple must be equal and opposite to each other.

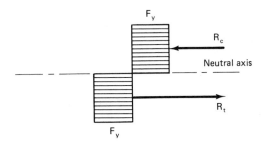

Figure 8-10 Stress diagram—full yield.

That is, the internal resultant of the compressive stress (R_c) must be equal to the internal resultant of the tensile stress (R_t). Therefore,

$$R_c = R_t$$

and

$$F_y A_c = F_y A_t$$

where F_y = yield stress
A_c = area on the compression side of the cross section
A_t = area on the tension side of the cross section

Since F_y is a constant value, $A_c = A_t$. This tells us that in a section stressed fully to yield, the neutral axis must be located at midarea.

It should be recalled that in elastic theory, the neutral axis is located at the point where the first moment of area $(A\bar{x})$ to either side of the axis is equal. We can see that the work necessary to locate the neutral axis by plastic theory is considerably simpler, especially if the cross section is unsymmetrical. For symmetrical cross sections, of course, the neutral axis is located at the midheight of the section regardless of the design method being used.

Determination of the Moment-Resisting Capacity

It was shown in the preceding discussion that based on the laws of static equilibrium,

$$R_c = R_t$$

and

$$F_y A_c = F_y A_t t$$

Referring now to Figure 8-11, the plastic moment (M_p) is

$$M_p = F_y A_c \bar{x}_c + F_y A_t \bar{x}_t$$

where $\bar{x}_c$ = distance from the centroid of the compression area to the neutral axis
$\bar{x}_t$ = distance from the centroid of the tension area to the neutral axis

It may be recalled from basic studies that the product of $A\bar{x}$ is known as the *first moment of area* $= Q$. Therefore,

$$M_p = F_y(Q_c + Q_t)$$

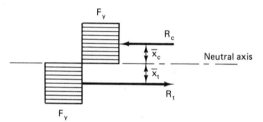

Figure 8-11 Resisting moment capacity.

where

$$Q_c + Q_t = Z$$

and Z is known as the *plastic modulus*. Therefore,

$$M_p = F_y Z$$

This is the basic bending stress equation for plastic design. The equation is structured in much the same manner as the bending stress equation used in allowable stress design, which is

$$M_a = F_a S$$

where M_a = allowable bending moment
F_a = allowable bending stress
S = elastic section modulus

Determination of the Load Factor for Plastic Design

Plastic design is a method where the design of a structural member is based on the maximum load that will produce "failure" (the collapse mechanism). The maximum, or ultimate, load that is used in the plastic design process is based on the anticipated dead and live loads, which are hypothetically inflated by a load factor and then viewed as ultimate loads. The anticipated dead and live loads will be referred to hereafter as the *service loads*. It is under the ultimate loads that the member will be designed to fail.

The load factor for dead and live loads recommended by the AISC Specification is 1.70. The load factor recommended when wind or earthquake forces are acting is 1.30 for the combined dead, live, and wind or earthquake forces. The rationale for this factor of 1.70 is based on the factor of safety being taken as the ratio of M_p:M_a. Therefore,

$$\frac{M_p}{M_a} = \frac{F_y}{F_a} \frac{Z}{S}$$

The ratio of Z/S is called the *shape factor*. The average shape factors for wide flange beams is approximately 1.13. Therefore,

$$\frac{F_y Z}{F_a S} = \frac{F_y}{0.66 F_y} (1.13) = (1.5)(1.13) = 1.7$$

Note that the expression $0.66F_y$ is the recommended allowable bending stress given by the AISC Specification.

Based on the manner in which the load factor is developed, and that only one plastic hinge needs to form for failure to occur, it should be clear that plastic design serves no purpose when designing simply supported beams. However, as we shall see shortly, where two or more plastic hinges must form for the collapse mechanism to occur, such as in a continuous beam, plastic design will be advantageous both from the standpoint of computational ease and, generally, saving of material.

APPLICATION OF THE EXPRESSIONS

To bring together all of the ideas of the preceding discussions, it seems that it would be useful now to look at several examples. We limit our considerations to the design of wide-flange beams of constant cross section. To do this, we will make reference to Data Sheets A15 to A20 in the Appendix throughout the design examples.

Example 8-1

Design the two-span continuous beam (select the most economical wide flange section) shown in Figure 8-12a. The beam is to be of A36 steel (F_y = 36 ksi). The service dead and live loads are given.

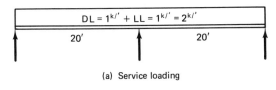

(a) Service loading

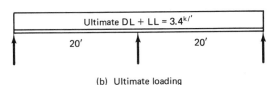

(b) Ultimate loading   **Figure 8-12**

Solution   To design this continuous member by the plastic design method, we must first inflate the given service load to the ultimate load by applying the recommended load factor of 1.70. The beam, with the ultimate load in place, is shown in Figure 8-12b. We will now proceed to select a wide flange section which will "fail" under this ultimate load (sometimes referred to as the *plastic limit load*).

To determine the maximum value of M_p, we isolate each span and view it as an independent three-hinged collapse mechanism. In this case both spans and loads are precisely the same. Isolating one of these spans, as shown in Figure 8-13, we have a member with one end restrained and one end pinned. This represents two hinges, a real hinge and a plastic hinge. The third hinge (a plastic

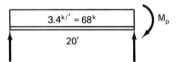

Figure 8-13   Isolated span.

hinge) will form at the point of maximum positive moment. Referring to Data Sheet A3, we have data for this condition. This expression yields

$$M_p = \frac{WL}{11.65} = \frac{(68 \text{ kips})(20 \text{ ft})}{11.65} = 116.7 \text{ ft-kips}$$

Now, referring to the Plastic Modulus Selection Tables in the Appendix (Data Sheets A15 to A20), we will select the most economical wide-flange section. The value of the plastic modulus *(Z)* required based on $F_y = 36$ ksi is

$$Z = \frac{M_p}{F_y} = \frac{(116.7 \text{ ft-kips})(12 \text{ in./ft})}{36} = 38.9 \text{ in.}^3$$

In scanning the tables we see that the lightest section to satisfy this requirement is a W14 × 26. The design is now complete.

   It is interesting to note that by the allowable stress design method, the required member would be a W18 × 35. The student can verify this by going through the design process using elastic theory and the service loads. Based on this, the member designed plastically represents a savings in material of 9 lb/ft or a total of 360 lb for this 40-ft-long beam.

Example 8-2

   In this case we have a two-span continuous beam with different spans and loading patterns, as shown in Figure 8-14a. Design a wide-flange section by plastic design.
   Solution   Inflate the given service loads by a factor of 1.70. The ultimate loads for which we will design are shown in Figure 8-14b. In this case we must investigate each span to see which will require the greater moment capacity. We do this by isolating each span as an independent mechanism and determining the value of M_p for each. These are shown in Figure 8-15. Referring to the plastic moment data in the Appendix (Data Sheets A15 to A20), we see that we have general expressions for each of these conditions.

Span 1: $$M_p = \frac{WL}{11.65} = \frac{(81.6 \text{ kips})(24 \text{ ft})}{11.65} = 168.1 \text{ ft-kips}$$

Span 2: $$M_p = \frac{PL}{4} = \frac{(17 \text{ kips})(30 \text{ ft})}{4} = 127.5 \text{ kips}$$

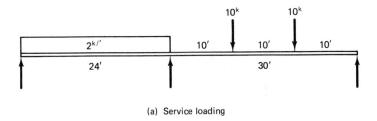

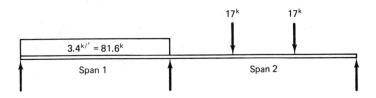

(a) Service loading

(b) Ultimate loading

Figure 8–14

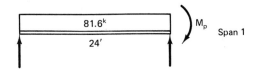

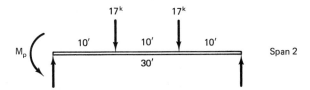

Figure 8–15

The condition of span 1 governs the design. Therefore, with $F_y = 36$ ksi,

$$Z = \frac{M_p}{F_y} = \frac{(168.1 \text{ ft-kips})(12 \text{ in./ft})}{36 \text{ ksi}} = 56 \text{ in.}^3$$

Scanning the Plastic Modulus Tables (Data Sheets A15 to A20), we find that the lightest section is a W18 × 35. It is recommended that an elastic analysis be made, using the service loads, for the sake of comparison.

In this case the selected member is overdesigned for span 2, and a plastic hinge will not form at the point of maximum positive moment. This means that there will still be elastic curvature in this span. In general, the deformed shape of this two-span continuous beam will be as shown in Figure 8–16.

169

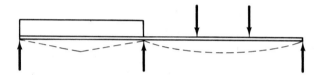

Figure 8-16

Example 8-3

In this case we wish to design a constant-section wide flange to satisfy the requirements of the three-span continuous beam shown in Figure 8-17a.

Solution   Again, we must inflate the given service loads by a factor of 1.70, and the ultimate loads for which we will design are shown in Figure 8-17b. We must now isolate each span as an independent mechanism and determine the

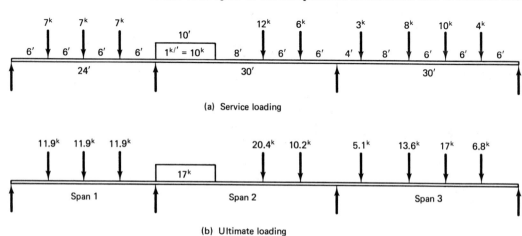

(a) Service loading

(b) Ultimate loading

Figure 8-17

largest plastic moment (M_p) capacity required. Let us first deal with span 1, which is shown in Figure 8-18. Referring to Data Sheets A15 to A20, we see that there is a general expression for this loading condition. Therefore:

$$Span\ 1: \quad M_p = \frac{PL}{3} = \frac{(11.9\ kips)(24\ ft)}{3} = 95.2\ ft\text{-}kips$$

Span 2 is shown as an independent mechanism in Figure 8-19. This represents a loading condition for which we do not have a general expression. However, the determination of M_p is a relatively simple matter. It was shown early

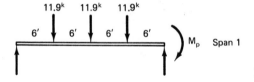

Figure 8-18

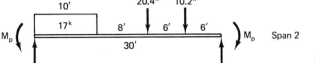

Figure 8–19

in this chapter that the plastic moment value for a member with both ends restrained is one-half of the moment as a simply supported beam. Therefore, let
us take span 2 as a simply supported beam and determine the moment for this
condition. The maximum simple span moment and its location (which will be
the location of the positive M_p) may be determined from the shear diagram.
This procedure is shown in Figure 8–20. From the area under the shear diagram
we find that simple span moment is 217.2 ft-kips. Therefore:

Span 2:
$$M_p = \frac{217.2 \text{ ft-kips}}{2} = 108.6 \text{ kips}$$

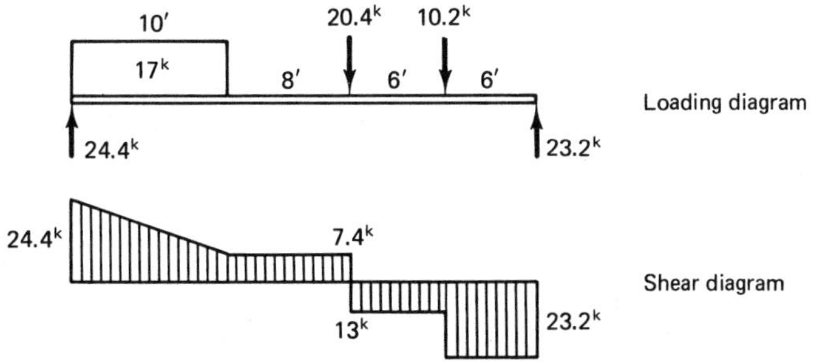

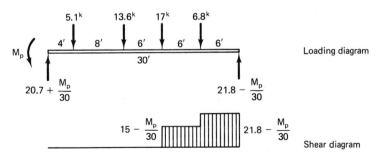

Figure 8–20

We must now investigate span 3, which is shown in Figure 8–21. This is
an irregular loading pattern for which we have no readily available general
expression. To determine the value of M_p and the location of the positive M_p,
we will use the same technique that was used earlier in this chapter to determine the value of M_p when only one end is restrained. This requires an assump-

Figure 8–21

tion regarding the location of the point where the shear diagram crosses the baseline, which, of course, is the location of the positive M_p. The load and shear diagrams are shown in Figure 8–21. It is assumed that the shear diagram will cross the baseline under the 17-kip load. Compute the area under the shear diagram to the right of the crossing:

Span 3:
$$M_p = \left(21.8 - \frac{M_p}{30}\right)(6) + \left(15 - \frac{M_p}{30}\right)(6)$$

$$= 220.8 - \frac{2M_p}{5}$$

$$= 157.7 \text{ ft-kips}$$

Using the value of M_p in the right-hand reaction verifies that the shear diagram does cross the baseline at the 17-kip load. Therefore, span 3 has the larger requirement of the three spans. Using A36 steel gives us

$$Z = \frac{M_p}{F_y} = \frac{(157.7 \text{ ft-kips})(12 \text{ in./ft})}{36 \text{ ksi}} = 52.6 \text{ in.}^3$$

Scanning the Plastic Modulus Tables in the Appendix, the lightest wide-flange section that will satisfy the requirements is a W16 × 31.

Unsymmetrical Sections

Although we are concerned primarily with the analysis and design of wide-flange sections, there are occasions when we must analyze the moment carrying capacity of an unsymmetrical cross section. This may occur, for example, where an existing wide-flange beam must be reinforced with additional steel plates in order to carry a greater load than that for which it was originally designed. This may happen where an existing structure frame is being used and the building is being rehabilitated for a new use.

We have, earlier in this chapter, developed the necessary principles and expressions that are needed for the analysis of an unsymmetrical section. We will therefore proceed immediately to an example.

Example 8–4

Determine the plastic moment (M_p) capacity of the section shown in Figure 8–22a, using A36 steel $(F_y = 36 \text{ ksi})$.

Solution   It was shown earlier in the chapter that the neutral axis in plastic analysis must be located at the *midarea* of the cross section. Therefore, in this case, the neutral axis is located at the bottom of the flange, as shown in Figure 8–22b. The stress diagram for the yielded section is shown in Figure 8–22c. Using the bending stress equation for plastic design:

$$M_p = F_y Z$$

where Z = the plastic modulus = $(Q_c + Q_t)$ and $Q_c + Q_t = A_c \bar{x}_c + A_t \bar{x}_t$.

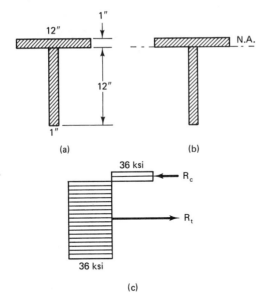

Figure 8–22

We must now determine the value of the plastic modulus *(Z)*:

$$Z = A_c \bar{x}_c + A_t \bar{x}_t = (12 \text{ in.}^2)(0.5 \text{ in.}) + (12 \text{ in.}^2)(6 \text{ in.}) = 78 \text{ in.}^3$$

Therefore,

$$M_p = F_y Z = \frac{(36 \text{ ksi})(78 \text{ in.}^3)}{12 \text{ in./ft}} = 234 \text{ ft-kips}$$

It is recommended that the reader determine the moment capacity of the section based on the allowable stress method. This should strengthen the understanding of the differences in the methods.

DEFLECTION CONSIDERATIONS

In Chapter 4 consideration for deflections was indicated as an important concern. Such concerns become even more important when a continuous beam is designed by the plastic design method. As a general rule the plastic design procedure will yield a smaller beam size than that which would be indicated by the allowable stress design method. Because of this, it is especially important that some judgments be made regarding the tolerable deflection and the actual amount that would occur in a given situation. We would be concerned with evaluating the deflection of a plastically designed beam under service loads. Under these loads the beam would, presumably, be behaving elastically. It seems of little importance to be concerned with the amount of deflection that would take place when the member fails as a three-hinged collapse mechanism. Consequently, the evaluation of deflection should be made under the service loads and by elastic theory procedures.

By and large, the plastic design method requires much less computation for the design of a continuous beam than does the allowable stress method. We would be defeating the purpose of computational ease if we designed the beam by plastic methods and then did a series of tedious computations by elastic methods in order to determine the precise value of deflection that would take place under service loadings. In response to this issue we can make certain very simple and quick judgments about the amount of deflection that would occur. Although these may be approximations (based on sound judgment), this would suffice under normal circumstances since the amount of tolerable deflection is, in itself, largely a matter of judgment. Primarily, we would be concerned with having a reasonable "ballpark" figure for the amount of deflection that would occur under service loads. To demonstrate some suggestions for the manner in which deflections may be estimated, let us review Example 8–1. This example considered the design of a two-span continuous beam.

To determine the approximate deflection we will use only the service load since we are concerned primarily with deflection at this level of loading. Figure 8–23 shows the two-span continuous beam with the service loading. To begin, let us isolate one of these spans and consider it as a simply supported beam. For this condition we can easily determine the deflection using Data Sheet A2. With the modulus of elasticity of steel taken as $E = 29,000$ ksi, the deflection as a simply supported beam would be*

$$\Delta = \frac{5WL^3}{384EI} = \frac{5(40 \text{ kips})(20 \text{ ft} \times 12 \text{ in./ft})}{384(29,000 \text{ ksi})(245 \text{ in.}^4)} = 1 \text{ in.}$$

This amount of deflection may seem excessive for a floor beam, but do not forget that this does not include the upward deflection due to the negative moment at the support. This will reduce the deflection that was determined as a simply supported beam. If necessary, a reasonable estimate of this reduction can be computed.

Service DL + LL = $2^{k/'}$

20'   20'

Figure 8–23   Two-span continuous beam.

To determine the effect on deflection of the negative moment at the support, we must have a value for the moment based on elastic behavior. To avoid the tedium of an elastic analysis under service loads, we can make some judgment about a value for the negative moment. This can be done by considering the *factor of safety* concept, which, for this purpose, will be defined as the ratio of M_p/M_e, where M_e is the moment based on elastic analysis under service loads. For the problem we are

*Remember: The member selected was a W14 × 26. The moment of inertia for this member is 245 in.^{4}.

considering, an elastic analysis gives a negative moment of 100 ft-kips and a positive moment of 56.25 ft-kips. The factor of safety for the negative moment is

$$\text{F.S.} = \frac{M_p}{-M_e} = \frac{116.7 \text{ ft-kips}}{100 \text{ ft-kips}} = 1.17$$

which is not the load factor of 1.70, as one might initially suspect. Let us determine the factor of safety for the positive moment and then discuss the matter a bit further.

$$\text{F.S.} = \frac{M_p}{+M_e} = \frac{116.7 \text{ ft-kips}}{56.25 \text{ ft-kips}} = 2.07$$

Again, this is different from the load factor of 1.70. It may be said that the factor of safety against *plastic collapse* (the formation of the three-hinged collapse mechanism) for the span being considered is 2.07.

For the purpose of estimating deflection under service loads we would be interested in having an estimate of the negative moment without going through an elastic analysis. It was shown in the preceding discussion that the factor of safety for the negative moment was less than the load factor. In most cases with reasonably normal spans and loading patterns, the negative moments by elastic analysis will be critical. In such cases the factor of safety against a plastic hinge formation will be less than the load factor. Having said this, it seems that it would be sensible simply to use the load factor of 1.70 to estimate the negative elastic moment. There may be a large percentage of error involved, but it would be on the safe side and, after all, we are only interested in making a judgment about the deflection situation. Therefore, for the case being considered,

$$\text{M} = \frac{M_p}{1.7} = \frac{116.7 \text{ ft-kips}}{1.7} = 70 \text{ ft-kips}$$

The effect of this negative moment is illustrated in Figure 8–24. This upward deflection at the midspan due to this moment is

$$\triangle = \frac{ML^2}{16EI} = \frac{(70 \text{ ft-kips})(12 \text{ in./ft})(20 \times 12)^2}{16(29,000 \text{ ksi})(245 \text{ in.}^4)} = 0.43 \text{ in.}$$

This amount of deflection subtracts from the deflection found for the simple beam condition, and this, although not a completely accurate figure, suggests that we will have a deflection somewhere in the range of about 0.60 in. For normal conditions this would probably be well within the range of acceptability.

It should be noted that the deflections computed in the preceding demonstration were based on midspan deflections. It should be noted that similar procedures

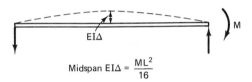

$$\text{Midspan } EI\triangle = \frac{ML^2}{16}$$

Figure 8–24 Midspan deflection due to end moment.

for determining deflections for continuous beams were presented in Chapter 4, where we designed beams by the allowable stress design method. In essence, the only difference in the process in this chapter has to do with making an estimate of the negative moment at the support due to service loads. As a further aid for determining deflections, the AISC *Manual of Steel Construction* has a number of general expressions for deflections for two-, three-, and four-span continuous beams which may be helpful for making appropriate judgments, although the variety of loading conditions is limited. These are shown in Section 2 of the *Manual*.

Should it be necessary to have *absolute* accuracy in the amount of deflection taking place under service loads in a plastically designed beam, it would be necessary to go through the tedium of a complete elastic analysis to determine the negative moments and, consequently, the location and magnitude of the maximum deflection. It is difficult, however, to imagine the need, in the great majority of circumstances, for such absolute accuracy.

LOAD AND RESISTANCE FACTOR DESIGN

A relatively new approach for the design of steel structures is known as load and resistance factor design (LRFD). LRFD is similar to the plastic design method in the sense that it is based on ultimate strength concepts (often referred to as "limit state" design). The American Institute of Steel Construction has a published specification for LRFD. In coming years it will be a feasible alternative to the now prevalent approach of allowable stress design.

Although plastic design has not generally been accepted by the profession in the United States, it is believed that the LRFD approach will be accepted in time. LRFD is intended to provide uniform reliability and predictability for steel structures and individual members subjected to a variety of possible loading conditions. Through the LRFD approach, structures are proportioned so that the ultimate strength (or *limit state*) is not exceeded when the structure or individual members are subjected to all possible load combinations.

In the allowable stress design procedure we use the actual values of working, or anticipated, loads with a factor of safety applied to the resistance, which, in essence, is the allowable stress for the condition to which the member is being subjected (i.e., flexure, tension, compression, etc.).

In the plastic design approach, studied earlier in this chapter, we design a structure based on the yield stress of the material as the prescribed resistance. The AISC Specification provides only two load factors with which to carry out the design procedure. They are 1.70 times the anticipated live and dead loads (as used in this chapter), and 1.3 times the live and dead loads acting in conjunction with wind or earthquake forces.

The LRFD approach uses a variety of load factors based on probability of occurrence and the likely magnitude when various types of loads act in conjunction with each other. In addition, LRFD uses variable factors for the resistance, depending on a variety of conditions.

In general, it may be said that LRFD is an approach that recognizes the degree of uncertainty of various types of loads and combinations of loads, as well as the degree of accuracy of the predicted strength (or resistance) of the structure.

CONCLUSION

In closing, it must be emphasized that only the basic ideas of plastic theory have been presented in this chapter. The purpose, as mentioned in the opening of this chapter, is to provide the reader with a vocabulary and a sense of the reserve strength of steel. We have not specifically discussed provisions of the AISC Specification for plastically designed beams, which provides us with requirements for shear, lateral stability, and so on. The reader who wishes to design by the plastic method should refer to Section 2 of the Specification.

This chapter should provide the student of architecture or related disciplines with the means to make quick preliminary estimates for continuous beams without the tedium required for an elastic analysis. This is often quite useful during the preliminary design phase, before a detailed engineering analysis is made.

SUPPLEMENTARY REFERENCES

BAKER, J. F., HORNE, M. R., and HEYMAN, J. *The Steel Skeleton,* Vol. II: *Plastic Behavior and Design.* Cambridge: Cambridge University Press, 1956.

BEEDLE, LYNN S. *Plastic Design of Steel Frames.* New York: John Wiley & Sons, Inc., 1958.

BENNETT, WILLIAM A. "Plastic Design of a 14-Story Apartment Building," *AISC Engineering Journal,* Vol. 4, No. 2, April 1967, pp. 68–72.

DRISCOLL, GEORGE C., Jr., et al. *Plastic Design of Multi-story Frames.* Bethlehem, Pa.: Lehigh University, 1965.

Guide to Load and Resistance Factor Design of Structural Steel Buildings. Chicago: American Institute of Steel Construction, 1986.

Load and Resistance Factor Design Manual. Chicago: American Institute of Steel Construction, 1986.

LU, LE-WU. "Design of Braced Multi-story Frames by the Plastic Design Method," *AISC Engineering Journal,* Vol. 4, No. 1, January 1967, pp. 1–9.

McCORMAC, JACK C. *Structural Steel Design.* Chapter 21, "Plastic Analysis;" Chapter 22, "Plastic Analysis and Design." Scranton, Pa.: International Textbook Company—College Division, 1967, pp. 487–537.

WILLIAMS, JAMES B., and GALAMBOS, THEODORE V. "Economic Study of a Braced Multi-story Steel Frame," *AISC Engineering Journal,* Vol. 5, No. 1, January 1968, pp. 2–11.

PROBLEMS

8.1. *Allowable stress and plastic design of continuous beams* (A36 steel, AISC Specification). F_a = 24 ksi, F_y = 36 ksi.

   (a) Given: the elastic analysis of maximum possible negative and positive moments based on critical arrangements of dead load and live load, as shown below. Determine the most economical wide-flange beams, based on the allowable stress design method.

   (b) Plastic design for loading = 1.70 (service dead and live loads). Determine the value of M_p for the critical span and select the most economical wide flange beam based on the plastic design method.

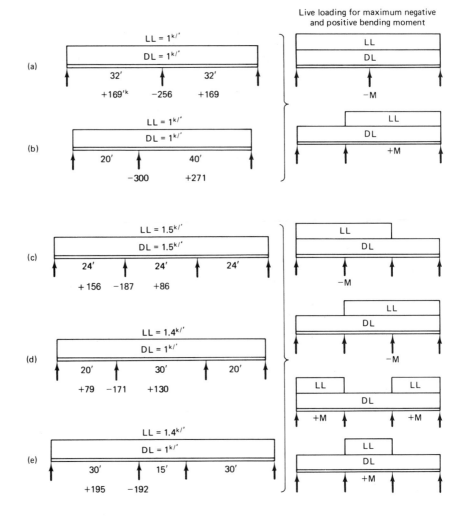

8.2. *Plastic design of continuous beams, constant section* (A36 steel, AISC Specification). The loads shown on each of the following are total (dead plus live) service loads. Using a load factor of 1.70 applied to the given service loads, determine the most economical wide-flange beam for each case, based on the plastic design method. Consider all beams to be appropriately laterally supported.

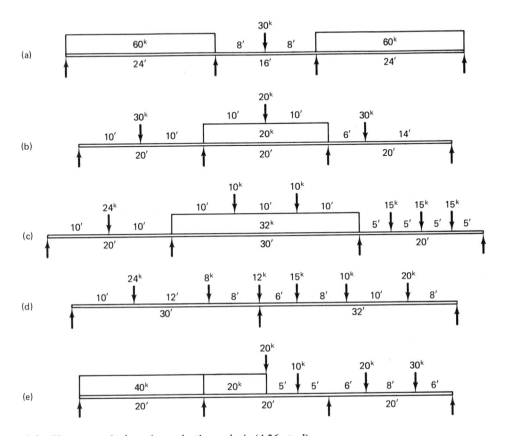

8.3. *Unsymmetrical section, plastic analysis* (A36 steel)

   (1) Determine the location of the plastic neutral axis and the value of the plastic modulus *(Z)*, for the section shown.

   (2) Determine the maximum values for the plastic limit loads W_p and P_p for the beam shown.

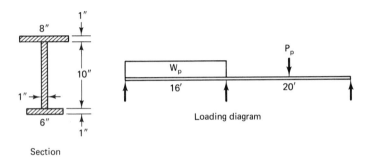

Section   Loading diagram

8.4. (1) Locate the plastic neutral axis and determine the value of the plastic modulus for the section shown.

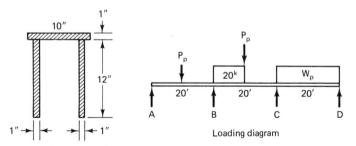

Loading diagram

(2) Determine the maximum values for the plastic limit loads P_p and W_p shown on spans A-B and C-D.

(3) Determine the maximum value of P_p, for span B-C, that may be added to the partial uniform load shown.

8.5. (1) Determine the value of M_p for the section shown, and the value of P_p that may be added to the given uniform load.

(2) Two plastic hinges will form at the restrained ends. Where will the third hinge form?

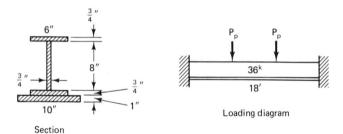

Section

Loading diagram

GLOSSARY

Allowable stress design A method of design and analysis based on an allowable stress and actual values of anticipated (or working) loads. The allowable stress is less than the yield stress of the steel. Sometimes referred to as elastic design or working stress design.

Base plate A plate at the base of a column that is anchored to the foundation. It is necessary to distribute the concentrated column load over a large area.

Bearing plate A steel plate under the end of a beam used to distribute the end reaction where the beam rests on a masonry or concrete support.

Bent A rigid beam and column assembly. A term most frequently used to describe a rigid frame.

Biaxial bending Bending of a structural member in both directions.

Box beam (or box girder) A hollow rectangular cross section, usually with diaphragms at regular intervals for stiffening.

Buckling load The load under which a compression member will deflect from the straight position.

Compact section A section that can develop its plastic moment capacity without local buckling.

Composite beam A steel beam carrying a concrete slab where the two are connected in a way to form an integral structural member.

Deformation The change in shape or dimensions of a structural element due to the application of stress.

Diaphragm A plate in a box beam or girder used to stiffen and otherwise maintain the original shape of the cross section.

Drift The lateral movement of a building due to wind.

Ductility The ability of a material to undergo large deformation without fracture.

Effective length A term usually applied to columns. It is the distance between the points of contraflexure (zero moment) of the anticipated buckled shape.

Elastic design *See* Allowable stress design.

Elasticity A property of a material that is capable of returning to its original shape and dimensions when a stress is removed.

Elastic limit The level of stress beyond which a material will not completely regain its original proportions and configuration when the stress is removed.

Erection plan A drawing usually prepared by a steel fabricator, showing the location and orientation of the steel members of a building frame.

Fabrication The process whereby steel members are cut to length, drilled, punched, and otherwise prepared for erection.

Factor of safety In general, the ratio of the yield stress of the steel to the allowable stress.

Fatigue A type of failure usually associated with repetitive stress conditions such as stress reversals and cyclic loading.

Girder Essentially, a beam that supports other beams.

Isotropic A material that has the same physical properties in all directions.

Joist A term usually associated with very small beams spaced very closely.

Lateral buckling Buckling of a member resulting in a lateral deflection or twist. Sometimes called lateral torsion.

Load factor A factor, used in plastic design, by which anticipated loads are multiplied to determine the ultimate loads to be carried.

Local buckling The buckling of part of a structural member which may result in total failure.

Malleable The ability to be rolled into a variety of shapes without breaking.

Mild steel Malleable steel with carbon content not greater than 0.30%. It is the type of steel used in architectural structures.

Modulus of elasticity The ratio of stress to strain of a material, within the elastic range. Structural steel has a modulus of elasticity *(E)* of 29,000 ksi.

Moment of inertia A geometric property of a cross section required to determine deflection and stresses due to bending.

Negative bending moment Bending moment that produces compression on the bottom side of a beam, and tension on the top side.

Nominal depth The notation used for the depth of a rolled steel member based on the set of rollers used at the steel mill. The actual depth is usually slightly different than the nominal depth.

Plastic axis The axis in a plastically designed cross section where the strains and stresses change from tensile to compressive. In plastic design the axis is located at the midarea of the cross section.

Plastic deformation Permanent, nonrecoverable deformation of steel stressed into the plastic range.

Plastic design A method whereby anticipated loads are multiplied by a load factor in order to determine the ultimate loads. The member is then designed to fail under these inflated loads.

Plastic hinge A zone in a steel member where the entire cross section has yielded. It is produced by the ultimate load.

Plastic modulus The resistance of a cross section that has completely yielded, based on the geometry of the section. Denoted by *Z*.

Plastic moment The moment that causes a section to yield, creating a plastic hinge.

Plastic range The range on the stress versus strain diagram where large strains occur without significant increase in stresses. The steel stress is at yield.

Plastification The penetration of yield stress from the outer fibers of a cross section toward the neutral axis. Plastification is said to be complete when the plastic moment is reached and a plastic hinge developed.

Plate girder Usually, a very deep beam made up of steel plates to form a wide-flange shape. Normally used where the size of a required structural girder exceeds that which is available in standard rolled sections.

Portal frame A rigidly jointed column and beam assembly. A "bent."

Positive bending moment Bending moment that produces tension on the bottom side of a beam, and compression on the top side.

Purlin Most commonly, the name given to members spanning between panel points of trusses.

Residual stress Permanent stresses introduced to a member due to rolling, uneven cooling, and so on.

Rigid frame A frame where the original angle between the vertical and horizontal member is maintained when the frame is loaded and deforms (within the elastic range of the material).

Section modulus The moment of inertia of a cross section divided by the distance from the neutral axis to the outermost fibers. Denoted by S. Sometimes referred to as the elastic section modulus.

Service load The actual load to which a structure will be subjected. Sometimes called the working load or anticipated load.

Shape factor The ratio of the plastic modulus of a section (Z) to the section modulus (S).

Shear stud A metal device, welded to the top flange of a steel beam, to resist shearing forces between the concrete slab and the steel beam. Used for composite beams.

Shop drawings Drawings prepared by the steel fabricator. They instruct the fabrication shop about the manner in which each piece of steel is to be cut, drilled, punched, and otherwise fabricated for erection in a building frame.

Sidesway Essentially, the same as drift, but specifically used in terms of the effect on columns in a tall building.

Skeletal structure A framework of steel beams and columns used to transmit all loads to the foundation of the building.

Strain Deformation of a material expressed as total deformation per unit of original length.

Stress versus strain diagram A diagram showing the relationship between stress and strain for a material. Data are based on laboratory testing.

Ultimate load The largest load a structure will support before failure (not necessarily collapse).

Unbraced length The distance between braced points of a structural member.

Web crippling The localized failure of the web of a beam or girder in the vicinity of a concentrated load or reaction.

Yield point The level of stress in a material beyond which large strains occur without significant (if any) increase in stress. It is less than the ultimate (breaking) stress.

Yield strain The strain at the yield point. Steel begins to show large amounts of plastic deformation.

Yield stress The stress measured at the yield point. Large deviation from proportionality between stress and strain.

APPENDIX

DATA SHEET A1 APPROXIMATE WEIGHTS OF BUILDING MATERIALS

Material	Weight (lb/ft²)	Material	Weight (lb/ft²)
Ceilings		Partitions	
Suspended system	1	Wood studs, 16 in. o.c.	2
Fiber tile	1	Steel studs, 16 in. o.c.	3
Lath and plaster	10	Gypsum board, 1/2 in.	2
Gypsum board, $\frac{1}{2}$ in.	2	Plaster	
		Gypsum, 1 in.	5
Floors		Cement, 1 in.	10
Steel deck	2–10	Lath (metal)	1
Reinf. conc., 1 in.			
Normal aggregate	12.5	Clay tile	
Lightweight aggregate	8–10	4 in.	18
		6 in.	28
Finishes		8 in.	34
Terrazzo, 1 in.	13		
Ceramic tile	10	Concrete block	*See* Walls
Quarry tile, $\frac{3}{4}$ in.	10	Brick	*See* Walls
Linoleum	1		
Hardwood, $\frac{7}{8}$ in.	4	Walls	
Sand fill, 1 in.	8	Concrete block	
		Normal aggregate	
Roofs		4 in.	30
Sheathing		6 in.	42
Wood, $\frac{3}{4}$ in.	3	8 in.	55
Gypsum, 1 in.	4	12 in.	80
		Lightweight aggregate	
Insulation		4 in.	20
Rigid, 1 in.	1.5	6 in.	28
Poured	2–3	8 in.	37
		12 in.	55
Copper or tin	1		
Felt and gravel		Brick	
3-ply	5.5	4 in.	40
5-ply	6	8 in.	80
Corrugated iron	3	12 in.	120
Corrugated asbestos-cement	4	Structural clay tile	
		4 in.	25
Clay tile	9–14	6 in.	30
Slate, $\frac{1}{4}$ in.	10	8 in.	33
		12 in.	45
		Stone, 4 in.	55
		Curtain walls	Obtain from manufacturer

DATA SHEET A2

MAXIMUM VALUES: SLOPE, DEFLECTION, AND BENDING MOMENT
Note: Values of slope and deflection to be divided by EI

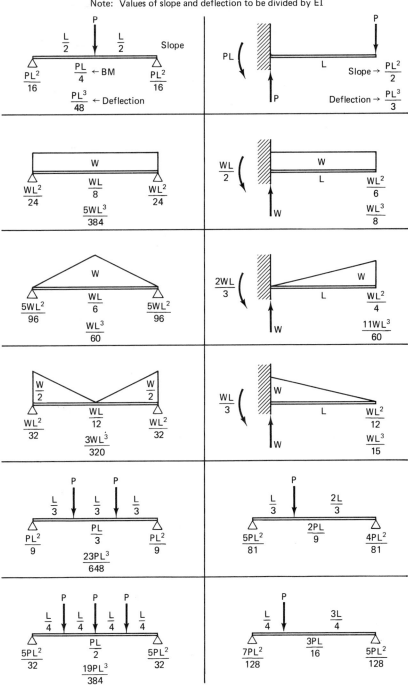

187

PLASTIC ANALYSIS
TYPICAL LOADINGS

RESTRAINED SPANS
CONSTANT SECTION

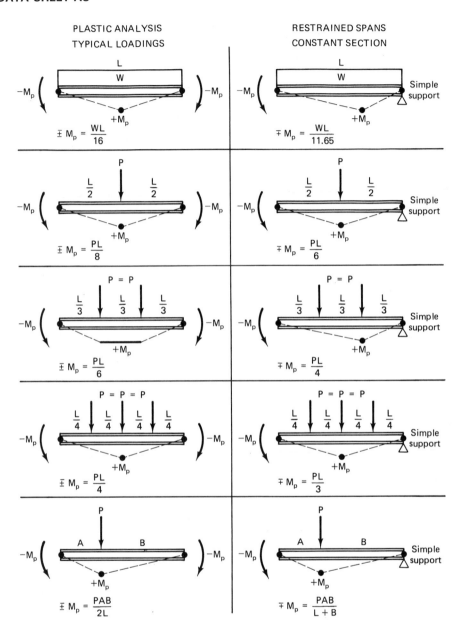

$\pm M_p = \dfrac{WL}{16}$

$\mp M_p = \dfrac{WL}{11.65}$

$\pm M_p = \dfrac{PL}{8}$

$\mp M_p = \dfrac{PL}{6}$

$\pm M_p = \dfrac{PL}{6}$

$\mp M_p = \dfrac{PL}{4}$

$\pm M_p = \dfrac{PL}{4}$

$\mp M_p = \dfrac{PL}{3}$

$\pm M_p = \dfrac{PAB}{2L}$

$\mp M_p = \dfrac{PAB}{L + B}$

1. The information given on Data Sheets A5 to A23 is taken from *Bethlehem Structural Shapes,* Catalog 3277c, 1983, and is reprinted here with the permission of Bethlehem Steel Corporation.

2. The data furnished on the aforementioned pages are provided to facilitate the learning process. From time to time, slight changes are made to this information. Therefore, if the reader is engaged in the actual design of a steel-framed structure, reference should be made to the most current data.

3. Data Sheets A5 to A11 contain the properties and dimensions for the design of structural steel members. Steel sections are designated according to:
   (a) Shape of the section
   (b) Nominal depth (in.) of the section
   (c) Weight (lb/lin ft) of the section
   For example, in the following designation:

$$W36 \times 300$$

   (a) The symbol "W" indicates that this is a wide-flange shape.
   (b) The number "36" indicates the nominal depth of the section. The actual depth will usually be slightly more or less than the nominal depth.
   (c) The number "300" indicates the weight of the section in terms of pounds per linear foot.

4. Data Sheets A12 and A13 give the elastic section moduli of wide-flange beams and they are arranged in a way to expedite the selection of the most economical beam (least weight) to satisfy design conditions. These will be used heavily in conjunction with the material of Chapter 4.

5. Data Sheet A14, the Moment of Inertia Selection Table, is arranged to expedite the selection of the most economical beam where deflection limitations govern the design of a beam.

6. Data Sheets A15 to A20 give the plastic section moduli for wide-flange sections. These data are arranged to expedite the selection of the most economical section when designing by the Plastic Design Method, which is presented in Chapter 8.

7. Data Sheets A21 to A23 provide information regarding the maximum unbraced length of the compression flange for members subjected to bending. These data used in conjunction with the material of Chapter 4.

8. Data Sheet A24 provides allowable stresses for compression members made of A36 steel. This information is used in conjunction with Chapter 5.

WIDE FLANGE SHAPES

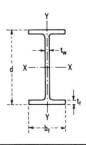

Theoretical Dimensions and Properties for **Designing**

Section Number	Weight per Foot	Area of Section	Depth of Section	Flange		Web Thick-ness	Axis X-X			Axis Y-Y			r_T
				Width	Thick-ness		I_x	S_x	r_x	I_y	S_y	r_y	
		A	d	b_f	t_f	t_w							
	lb	in.²	in.	in.	in.	in.	in.⁴	in.³	in.	in.⁴	in.³	in.	in.
W36 x 300	88.3	36.74	16.655	1.680	0.945	20300	1110	15.2	1300	156	3.83	4.39
280	82.4	36.52	16.595	1.570	0.885	18900	1030	15.1	1200	144	3.81	4.37
260	76.5	36.26	16.550	1.440	0.840	17300	953	15.0	1090	132	3.78	4.34
245	72.1	36.08	16.510	1.350	0.800	16100	895	15.0	1010	123	3.75	4.32
230	67.6	35.90	16.470	1.260	0.760	15000	837	14.9	940	114	3.73	4.30
W36 x 210	61.8	36.69	12.180	1.360	0.830	13200	719	14.6	411	67.5	2.58	3.09
194	57.0	36.49	12.115	1.260	0.765	12100	664	14.6	375	61.9	2.56	3.07
182	53.6	36.33	12.075	1.180	0.725	11300	623	14.5	347	57.6	2.55	3.05
170	50.0	36.17	12.030	1.100	0.680	10500	580	14.5	320	53.2	2.53	3.04
160	47.0	36.01	12.000	1.020	0.650	9750	542	14.4	295	49.1	2.50	3.02
150	44.2	35.85	11.975	0.940	0.625	9040	504	14.3	270	45.1	2.47	2.99
135	39.7	35.55	11.950	0.790	0.600	7800	439	14.0	225	37.7	2.38	2.93
W33 x 241	70.9	34.18	15.860	1.400	0.830	14200	829	14.1	932	118	3.63	4.17
221	65.0	33.93	15.805	1.275	0.775	12800	757	14.1	840	106	3.59	4.15
201	59.1	33.68	15.745	1.150	0.715	11500	684	14.0	749	95.2	3.56	4.12
W33 x 152	44.7	33.49	11.565	1.055	0.635	8160	487	13.5	273	47.2	2.47	2.94
141	41.6	33.30	11.535	0.960	0.605	7450	448	13.4	246	42.7	2.43	2.92
130	38.3	33.09	11.510	0.855	0.580	6710	406	13.2	218	37.9	2.39	2.88
118	34.7	32.86	11.480	0.740	0.550	5900	359	13.0	187	32.6	2.32	2.84
W30 x 211	62.0	30.94	15.105	1.315	0.775	10300	663	12.9	757	100	3.49	3.99
191	56.1	30.68	15.040	1.185	0.710	9170	598	12.8	673	89.5	3.46	3.97
173	50.8	30.44	14.985	1.065	0.655	8200	539	12.7	598	79.8	3.43	3.94
W30 x 132	38.9	30.31	10.545	1.000	0.615	5770	380	12.2	196	37.2	2.25	2.68
124	36.5	30.17	10.515	0.930	0.585	5360	355	12.1	181	34.4	2.23	2.66
116	34.2	30.01	10.495	0.850	0.565	4930	329	12.0	164	31.3	2.19	2.64
108	31.7	29.83	10.475	0.760	0.545	4470	299	11.9	146	27.9	2.15	2.61
99	29.1	29.65	10.450	0.670	0.520	3990	269	11.7	128	24.5	2.10	2.57

All shapes on these pages have parallel-faced flanges.

WIDE FLANGE SHAPES

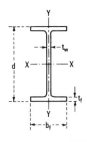

Theoretical Dimensions and Properties for **Designing**

Section Number	Weight per Foot	Area of Section	Depth of Section	Flange		Web Thick-ness	Axis X-X			Axis Y-Y			r_T
				Width	Thick-ness		I_x	S_x	r_x	I_y	S_y	r_y	
		A	d	b_f	t_f	t_w							
	lb	in.²	in.	in.	in.	in.	in.⁴	in.³	in.	in.⁴	in.³	in.	in.
W27 x **178**	52.3	27.81	14.085	1.190	0.725	6990	502	11.6	555	78.8	3.26	3.72
161	47.4	27.59	14.020	1.080	0.660	6280	455	11.5	497	70.9	3.24	3.70
146	42.9	27.38	13.965	0.975	0.605	5630	411	11.4	443	63.5	3.21	3.68
W27 x **114**	33.5	27.29	10.070	0.930	0.570	4090	299	11.0	159	31.5	2.18	2.58
102	30.0	27.09	10.015	0.830	0.515	3620	267	11.0	139	27.8	2.15	2.56
94	27.7	26.92	9.990	0.745	0.490	3270	243	10.9	124	24.8	2.12	2.53
84	24.8	26.71	9.960	0.640	0.460	2850	213	10.7	106	21.2	2.07	2.49
W24 x **162**	47.7	25.00	12.955	1.220	0.705	5170	414	10.4	443	68.4	3.05	3.45
146	43.0	24.74	12.900	1.090	0.650	4580	371	10.3	391	60.5	3.01	3.43
131	38.5	24.48	12.855	0.960	0.605	4020	329	10.2	340	53.0	2.97	3.40
117	34.4	24.26	12.800	0.850	0.550	3540	291	10.1	297	46.5	2.94	3.37
104	30 6	24.06	12.750	0.750	0.500	3100	258	10.1	259	40.7	2.91	3.35
W24 x **94**	27.7	24.31	9.065	0.875	0.515	2700	222	9.87	109	24.0	1.98	2.33
84	24.7	24.10	9.020	0.770	0.470	2370	196	9.79	94.4	20.9	1.95	2.31
76	22.4	23.92	8.990	0.680	0.440	2100	176	9.69	82.5	18.4	1.92	2.29
68	20.1	23.73	8.965	0.585	0.415	1830	154	9.55	70.4	15.7	1.87	2.26
W24 x **62**	18.2	23.74	7.040	0.590	0.430	1550	131	9.23	34.5	9.80	1.38	1.71
55	16.2	23.57	7.005	0.505	0.395	1350	114	9.11	29.1	8.30	1.34	1.68
W21 x **147**	43.2	22.06	12.510	1.150	0.720	3630	329	9.17	376	60.1	2.95	3.34
132	38.8	21.83	12.440	1.035	0.650	3220	295	9.12	333	53.5	2.93	3.31
122	35.9	21.68	12.390	0.960	0.600	2960	273	9.09	305	49.2	2.92	3.30
111	32.7	21.51	12.340	0.875	0.550	2670	249	9.05	274	44.5	2.90	3.28
101	29.8	21.36	12.290	0.800	0.500	2420	227	9.02	248	40.3	2.89	3.27
W21 x **93**	27.3	21.62	8.420	0.930	0.580	2070	192	8.70	92.9	22.1	1.84	2.17
83	24.3	21.43	8.355	0.835	0.515	1830	171	8.67	81.4	19.5	1.83	2.15
73	21.5	21.24	8.295	0.740	0.455	1600	151	8.64	70.6	17.0	1.81	2.13
68	20.0	21.13	8.270	0.685	0.430	1480	140	8.60	64.7	15.7	1.80	2.12
62	18.3	20.99	8.240	0.615	0.400	1330	127	8.54	57.5	13.9	1.77	2.10
W21 x **57**	16.7	21.06	6.555	0.650	0.405	1170	111	8.36	30.6	9.35	1.35	1.64
50	14.7	20.83	6.530	0.535	0.380	984	94.5	8.18	24.9	7.64	1.30	1.60
44	13.0	20.66	6.500	0.450	0.350	843	81.6	8.06	20.7	6.36	1.26	1.57

All shapes on these pages have parallel-faced flanges.

WIDE FLANGE SHAPES

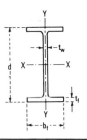

Theoretical Dimensions and Properties for **Designing**

Section Number	Weight per Foot	Area of Section	Depth of Section	Flange Width	Flange Thickness	Web Thickness	Axis X-X I_x	Axis X-X S_x	Axis X-X r_x	Axis Y-Y I_y	Axis Y-Y S_y	Axis Y-Y r_y	r_T
		A	d	b_f	t_f	t_w							
	lb	in.²	in.	in.	in.	in.	in.⁴	in.³	in.	in.⁴	in.³	in.	in.
W18 x 119	35.1	18.97	11.265	1.060	0.655	2190	231	7.90	253	44.9	2.69	3.02
106	31.1	18.73	11.200	0.940	0.590	1910	204	7.84	220	39.4	2.66	3.00
97	28.5	18.59	11.145	0.870	0.535	1750	188	7.82	201	36.1	2.65	2.99
86	25.3	18.39	11.090	0.770	0.480	1530	166	7.77	175	31.6	2.63	2.97
76	22.3	18.21	11.035	0.680	0.425	1330	146	7.73	152	27.6	2.61	2.95
W18 x 71	20.8	18.47	7.635	0.810	0.495	1170	127	7.50	60.3	15.8	1.70	1.98
65	19.1	18.35	7.590	0.750	0.450	1070	117	7.49	54.8	14.4	1.69	1.97
60	17.6	18.24	7.555	0.695	0.415	984	108	7.47	50.1	13.3	1.69	1.96
55	16.2	18.11	7.530	0.630	0.390	890	98.3	7.41	44.9	11.9	1.67	1.95
50	14.7	17.99	7.495	0.570	0.355	800	88.9	7.38	40.1	10.7	1.65	1.94
W18 x 46	13.5	18.06	6.060	0.605	0.360	712	78.8	7.25	22.5	7.43	1.29	1.54
40	11.8	17.90	6.015	0.525	0.315	612	68.4	7.21	19.1	6.35	1.27	1.52
35	10.3	17.70	6.000	0.425	0.300	510	57.6	7.04	15.3	5.12	1.22	1.49
W16 x 100	29.4	16.97	10.425	0.985	0.585	1490	175	7.10	186	35.7	2.52	2.81
89	26.2	16.75	10.365	0.875	0.525	1300	155	7.05	163	31.4	2.49	2.79
77	22.6	16.52	10.295	0.760	0.455	1110	134	7.00	138	26.9	2.47	2.77
67	19.7	16.33	10.235	0.665	0.395	954	117	6.96	119	23.2	2.46	2.75
W16 x 57	16.8	16.43	7.120	0.715	0.430	758	92.2	6.72	43.1	12.1	1.60	1.86
50	14.7	16.26	7.070	0.630	0.380	659	81.0	6.68	37.2	10.5	1.59	1.84
45	13.3	16.13	7.035	0.565	0.345	586	72.7	6.65	32.8	9.34	1.57	1.83
40	11.8	16.01	6.995	0.505	0.305	518	64.7	6.63	28.9	8.25	1.57	1.82
36	10.6	15.86	6.985	0.430	0.295	448	56.5	6.51	24.5	7.00	1.52	1.79
W16 x 31	9.12	15.88	5.525	0.440	0.275	375	47.2	6.41	12.4	4.49	1.17	1.39
26	7.68	15.69	5.500	0.345	0.250	301	38.4	6.26	9.59	3.49	1.12	1.36

All shapes on these pages have parallel-faced flanges.

WIDE FLANGE SHAPES

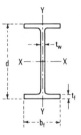

Theoretical Dimensions and Properties for **Designing**

Section Number	Weight per Foot	Area of Section A	Depth of Section d	Flange		Web Thick-ness t_w	Axis X-X			Axis Y-Y			r_T
				Width b_f	Thick-ness t_f		I_x	S_x	r_x	I_y	S_y	r_y	
	lb	in.²	in.	in.	in.	in.	in.⁴	in.³	in.	in.⁴	in.³	in.	in.
W14 x	730*	215	22.42	17.890	4.910	3.070	14300	1280	8.17	4720	527	4.69	4.99
	665*	196	21.64	17.650	4.520	2.830	12400	1150	7.98	4170	472	4.62	4.92
	605*	178	20.92	17.415	4.160	2.595	10800	1040	7.80	3680	423	4.55	4 85
	550*	162	20.24	17.200	3.820	2.380	9430	931	7.63	3250	378	4.49	4.79
	500*	147	19.60	17.010	3.500	2.190	8210	838	7.48	2880	339	4.43	4.73
	455*	134	19.02	16.835	3.210	2.015	7190	756	7.33	2560	304	4.38	4.68
W14 x	426	125	18.67	16.695	3.035	1.875	6600	707	7.26	2360	283	4.34	4.64
	398	117	18.29	16.590	2.845	1.770	6000	656	7.16	2170	262	4.31	4.61
	370	109	17.92	16.475	2.660	1.655	5440	607	7.07	1990	241	4.27	4.57
	342	101	17.54	16.360	2.470	1.540	4900	559	6.98	1810	221	4.24	4.54
	311	91.4	17.12	16.230	2.260	1.410	4330	506	6.88	1610	199	4.20	4.50
	283	83.3	16.74	16.110	2.070	1.290	3840	459	6.79	1440	179	4.17	4.46
	257	75.6	16.38	15.995	1.890	1.175	3400	415	6.71	1290	161	4.13	4.43
	233	68.5	16.04	15.890	1.720	1.070	3010	375	6.63	1150	145	4.10	4.40
	211	62.0	15.72	15.800	1.560	0.980	2660	338	6.55	1030	130	4.07	4.37
	193	56.8	15.48	15.710	1.440	0.890	2400	310	6.50	931	119	4.05	4.35
	176	51.8	15.22	15.650	1.310	0.830	2140	281	6.43	838	107	4.02	4.32
	159	46.7	14.98	15.565	1.190	0.745	1900	254	6.38	748	96.2	4.00	4.30
	145	42.7	14.78	15.500	1.090	0.680	1710	232	6.33	677	87.3	3.98	4.28
W14 x	132	38.8	14.66	14.725	1.030	0.645	1530	209	6.28	548	74.5	3.76	4.05
	120	35.3	14.48	14.670	0.940	0.590	1380	190	6.24	495	67.5	3.74	4.04
	109	32.0	14.32	14.605	0.860	0.525	1240	173	6.22	447	61.2	3.73	4.02
	99	29.1	14.16	14.565	0.780	0.485	1110	157	6.17	402	55.2	3.71	4.00
	90	26.5	14.02	14.520	0.710	0.440	999	143	6.14	362	49.9	3.70	3.99
W14 x	82	24.1	14.31	10.130	0.855	0.510	882	123	6.05	148	29.3	2.48	2.74
	74	21.8	14.17	10.070	0.785	0.450	796	112	6.04	134	26.6	2.48	2.72
	68	20.0	14.04	10.035	0.720	0.415	723	103	6.01	121	24.2	2.46	2.71
	61	17.9	13.89	9.995	0.645	0.375	640	92.2	5.98	107	21.5	2.45	2.70
W14 x	53	15.6	13.92	8.060	0.660	0.370	541	77.8	5.89	57.7	14.3	1.92	2.15
	48	14.1	13.79	8.030	0.595	0.340	485	70.3	5.85	51.4	12.8	1.91	2.13
	43	12.6	13.66	7.995	0.530	0.305	428	62.7	5.82	45.2	11.3	1.89	2.12

*These shapes have a 1°-00′ (1.75%) flange slope. Flange thicknesses shown are average thicknesses. Properties shown are for a parallel flange section.

All other shapes on these pages have parallel-faced flanges.

WIDE FLANGE SHAPES

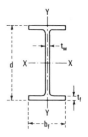

Theoretical Dimensions and Properties for **Designing**

Section Number	Weight per Foot	Area of Section	Depth of Section	Flange Width	Flange Thickness	Web Thickness	Axis X-X I_x	S_x	r_x	Axis Y-Y I_y	S_y	r_y	r_T
		A	d	b_f	t_f	t_w							
	lb	in.²	in.	in.	in.	in.	in.⁴	in.³	in.	in.⁴	in.³	in.	in.
W14 x 38		11.2	14.10	6.770	0.515	0.310	385	54.6	5.88	26.7	7.88	1.55	1.77
34		10.0	13.98	6.745	0.455	0.285	340	48.6	5.83	23.3	6.91	1.53	1.76
30		8.85	13.84	6.730	0.385	0.270	291	42.0	5.73	19.6	5.82	1.49	1.74
W14 x 26		7.69	13.91	5.025	0.420	0.255	245	35.3	5.65	8.91	3.54	1.08	1.28
22		6.49	13.74	5.000	0.335	0.230	199	29.0	5.54	7.00	2.80	1.04	1.25
W12 x 190		55.8	14.38	12.670	1.735	1.060	1890	263	5.82	589	93.0	3.25	3.50
170		50.0	14.03	12.570	1.560	0.960	1650	235	5.74	517	82.3	3.22	3.47
152		44.7	13.71	12.480	1.400	0.870	1430	209	5.66	454	72.8	3.19	3.44
136		39.9	13.41	12.400	1.250	0.790	1240	186	5.58	398	64.2	3.16	3.41
120		35.3	13.12	12.320	1.105	0.710	1070	163	5.51	345	56.0	3.13	3.38
106		31.2	12.89	12.220	0.990	0.610	933	145	5.47	301	49.3	3.11	3.36
96		28.2	12.71	12.160	0.900	0.550	833	131	5.44	270	44.4	3.09	3.34
87		25.6	12.53	12.125	0.810	0.515	740	118	5.38	241	39.7	3.07	3.32
79		23.2	12.38	12.080	0.735	0.470	662	107	5.34	216	35.8	3.05	3.31
72		21.1	12.25	12.040	0.670	0.430	597	97.4	5.31	195	32.4	3.04	3.29
65		19.1	12.12	12.000	0.605	0.390	533	87.9	5.28	174	29.1	3.02	3.28
W12 x 58		17.0	12.19	10.010	0.640	0.360	475	78.0	5.28	107	21.4	2.51	2.72
53		15.6	12.06	9.995	0.575	0.345	425	70.6	5.23	95.8	19.2	2.48	2.71
W12 x 50		14.7	12.19	8.080	0.640	0.370	394	64.7	5.18	56.3	13.9	1.96	2.17
45		13.2	12.06	8.045	0.575	0.335	350	58.1	5.15	50.0	12.4	1.94	2.15
40		11.8	11.94	8.005	0.515	0.295	310	51.9	5.13	44.1	11.0	1.93	2.14
W12 x 35		10.3	12.50	6.560	0.520	0.300	285	45.6	5.25	24.5	7.47	1.54	1.74
30		8.79	12.34	6.520	0.440	0.260	238	38.6	5.21	20.3	6.24	1.52	1.73
26		7.65	12.22	6.490	0.380	0.230	204	33.4	5.17	17.3	5.34	1.51	1.72
W12 x 22		6.48	12.31	4.030	0.425	0.260	156	25.4	4.91	4.66	2.31	0.848	1.02
19		5.57	12.16	4.005	0.350	0.235	130	21.3	4.82	3.76	1.88	0.822	0.997
16		4.71	11.99	3.990	0.265	0.220	103	17.1	4.67	2.82	1.41	0.773	0.963
14		4.16	11.91	3.970	0.225	0.200	88.6	14.9	4.62	2.36	1.19	0.753	0.946

All shapes on these pages have parallel-faced flanges.

WIDE FLANGE SHAPES

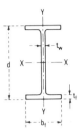

Theoretical Dimensions and Properties for **Designing**

Section Number	Weight per Foot	Area of Section	Depth of Section	Flange		Web Thick-ness	Axis X-X			Axis Y-Y			r_T
				Width	Thick-ness		I_x	S_x	r_x	I_y	S_y	r_y	
		A	d	b_f	t_f	t_w							
	lb	in.²	in.	in.	in.	in.	in.⁴	in.³	in.	in.⁴	in.³	in.	in.
W10 x 112	112	32.9	11.36	10.415	1.250	0.755	716	126	4.66	236	45.3	2.68	2.88
100	100	29.4	11.10	10.340	1.120	0.680	623	112	4.60	207	40.0	2.65	2.85
88	88	25.9	10.84	10.265	0.990	0.605	534	98.5	4.54	179	34.8	2.63	2.83
77	77	22.6	10.60	10.190	0.870	0.530	455	85.9	4.49	154	30.1	2.60	2.80
68	68	20.0	10.40	10.130	0.770	0.470	394	75.7	4.44	134	26.4	2.59	2.79
60	60	17.6	10.22	10.080	0.680	0.420	341	66.7	4.39	116	23.0	2.57	2.77
54	54	15.8	10.09	10.030	0.615	0.370	303	60.0	4.37	103	20.6	2.56	2.75
49	49	14.4	9.98	10.000	0.560	0.340	272	54.6	4.35	93.4	18.7	2.54	2.74
W10 x 45	45	13.3	10.10	8.020	0.620	0.350	248	49.1	4.33	53.4	13.3	2.01	2.18
39	39	11.5	9.92	7.985	0.530	0.315	209	42.1	4.27	45.0	11.3	1.98	2.16
33	33	9.71	9.73	7.960	0.435	0.290	170	35.0	4.19	36.6	9.20	1.94	2.14
W10 x 30	30	8.84	10.47	5.810	0.510	0.300	170	32.4	4.38	16.7	5.75	1.37	1.55
26	26	7.61	10.33	5.770	0.440	0.260	144	27.9	4.35	14.1	4.89	1.36	1.54
22	22	6.49	10.17	5.750	0.360	0.240	118	23.2	4.27	11.4	3.97	1.33	1.51
W10 x 19	19	5.62	10.24	4.020	0.395	0.250	96.3	18.8	4.14	4.29	2.14	0.874	1.03
17	17	4.99	10.11	4.010	0.330	0.240	81.9	16.2	4.05	3.56	1.78	0.845	1.01
15	15	4.41	9.99	4.000	0.270	0.230	68.9	13.8	3.95	2.89	1.45	0.810	0.987
12	12	3.54	9.87	3.960	0.210	0.190	53.8	10.9	3.90	2.18	1.10	0.785	0.965
W8 x 67	67	19.7	9.00	8.280	0.935	0.570	272	60.4	3.72	88.6	21.4	2.12	2.28
58	58	17.1	8.75	8.220	0.810	0.510	228	52.0	3.65	75.1	18.3	2.10	2.26
48	48	14.1	8.50	8.110	0.685	0.400	184	43.3	3.61	60.9	15.0	2.08	2.23
40	40	11.7	8.25	8.070	0.560	0.360	146	35.5	3.53	49.1	12.2	2.04	2.21
35	35	10.3	8.12	8.020	0.495	0.310	127	31.2	3.51	42.6	10.6	2.03	2.20
31	31	9.13	8.00	7.995	0.435	0.285	110	27.5	3.47	37.1	9.27	2.02	2.18
W8 x 28	28	8.25	8.06	6.535	0.465	0.285	98.0	24.3	3.45	21.7	6.63	1.62	1.77
24	24	7.08	7.93	6.495	0.400	0.245	82.8	20.9	3.42	18.3	5.63	1.61	1.76
W8 x 21	21	6.16	8.28	5.270	0.400	0.250	75.3	18.2	3.49	9.77	3.71	1.26	1.41
18	18	5.26	8.14	5.250	0.330	0.230	61.9	15.2	3.43	7.97	3.04	1.23	1.39
W8 x 15	15	4.44	8.11	4.015	0.315	0.245	48.0	11.8	3.29	3.41	1.70	0.876	1.03
13	13	3.84	7.99	4.000	0.255	0.230	39.6	9.91	3.21	2.73	1.37	0.843	1.01
10	10	2.96	7.89	3.940	0.205	0.170	30.8	7.81	3.22	2.09	1.06	0.841	0.994

All shapes on these pages have parallel-faced flanges.

WIDE FLANGE SHAPES

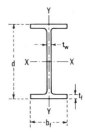

Theoretical Dimensions and Properties for **Designing**

Section Number	Weight per Foot	Area of Section	Depth of Section	Flange			Axis X-X			Axis Y-Y			
				Width	Thick-ness	Web Thick-ness	I_x	S_x	r_x	I_y	S_y	r_y	r_T
		A	d	b_f	t_f	t_w							
	lb	in.²	in.	in.	in.	in.	in.⁴	in.³	in.	in.⁴	in.³	in.	in.
W6 x 25	25	7.34	6.38	6.080	0.455	0.320	53.4	16.7	2.70	17.1	5.61	1.52	1.66
20	20	5.87	6.20	6.020	0.365	0.260	41.4	13.4	2.66	13.3	4.41	1.50	1.64
15	15	4.43	5.99	5.990	0.260	0.230	29.1	9.72	2.56	9.32	3.11	1.45	1.61
W6 x 16	16	4.74	6.28	4.030	0.405	0.260	32.1	10.2	2.60	4.43	2.20	0.967	1.08
12	12	3.55	6.03	4.000	0.280	0.230	22.1	7.31	2.49	2.99	1.50	0.918	1.05
9	9	2.68	5.90	3.940	0.215	0.170	16.4	5.56	2.47	2.20	1.11	0.905	1.03
W5 x 19	19	5.54	5.15	5.030	0.430	0.270	26.2	10.2	2.17	9.13	3.63	1.28	1.38
16	16	4.68	5.01	5.000	0.360	0.240	21.3	8.51	2.13	7.51	3.00	1.27	1.37
†W4 x 13	13	3.83	4.16	4.060	0.345	0.280	11.3	5.46	1.72	3.86	1.90	1.00	1.10

MISCELLANEOUS SHAPE

Theoretical Dimensions and Properties for **Designing**

Section Number	Weight per Foot	Area of Section	Depth of Section	Flange			Axis X-X			Axis Y-Y			
				Width	Thick-ness	Web Thick-ness	I_x	S_x	r_x	I_y	S_y	r_y	r_T
		A	d	b_f	t_f	t_w							
	lb	in.²	in.	in.	in.	in.	in.⁴	in.³	in.	in.⁴	in.³	in.	in.
†M5 x 18.9	18.9	5.55	5.00	5.003	0.416	0.316	24.1	9.63	2.08	7.86	3.14	1.19	1.32

†W4 x 13 and M5 x 18.9 have flange slopes of 2.0 and 7.4 pct respectively. Flange thickness shown for these sections are average thicknesses. Properties are the same as if flanges were parallel.

All other shapes on these pages have parallel-faced flanges.

ELASTIC SECTION MODULUS

These tables are in accordance with the AISC Specification for the Design, Fabrication & Erection of Structural Steel for Buildings, Supplement No. 3 (1974).

Sections shown in **bold face** are "Weight Economy Sections."

S_x in.³	Shape	F'_y ksi	S_x in.³	Shape	F'_y ksi	S_x in.³	Shape	F'_y ksi
1110	W36x300	**	439	W36x135	**	213	W27x84	**
			415	W14x257	**	209	W14x132	**
1030	W36x280	**	414	W24x162	**	209	W12x152	**
			411	W27x146	**	204	W18x106	**
953	W36x260	**						
			406	W33x130	**	196	W24x84	**
895	W36x245	**	380	W30x132	**	192	W21x93	**
			375	W14x233	**	190	W14x120	**
837	W36x230	**	371	W24x146	**	188	W18x97	**
829	W33x241	**				186	W12x136	**
			359	W33x118	**			
757	W33x221	**	355	W30x124	**	176	W24x76	**
			338	W14x211	**	175	W16x100	**
719	W36x210	**				173	W14x109	58.6
707	W14x426	**	329	W30x116	**	171	W21x83	**
			329	W24x131	**	166	W18x86	**
684	W33x201	**	329	W21x147	**	163	W12x120	**
			310	W14x193	**	157	W14x99	48.5
664	W36x194	**				155	W16x89	**
663	W30x211	**	299	W30x108	**			
656	W14x398	**	299	W27x114	**	154	W24x68	**
			295	W21x132	**	151	W21x73	**
623	W36x182	**	291	W24x117	**	146	W18x76	**
607	W14x370	**	281	W14x176	**	145	W12x106	**
598	W30x191	**	273	W21x122	**	143	W14x90	40.4
580	W36x170	**						
559	W14x342	**	269	W30x99	**	140	W21x68	**
			267	W27x102	**	134	W16x77	**
542	W36x160	**	263	W12x190	**			
539	W30x173	**	258	W24x104	58.5	131	W24x62	**
506	W14x311	**	254	W14x159	**	131	W12x96	**
			249	W21x111	**			
504	W36x150	**				127	W21x62	**
502	W27x178	**	243	W27x94	**	127	W18x71	**
487	W33x152	**	235	W12x170	**	126	W10x112	**
459	W14x283	**	232	W14x145	**	123	W14x82	**
455	W27x161	**	231	W18x119	**	118	W12x87	**
			227	W21x101	**	117	W18x65	**
448	W33x141	**				117	W16x67	**
			222	W24x94	**			

**Theoretical maximum yield stress exceeds 60 ksi.

ELASTIC SECTION MODULUS

These tables are in accordance with the AISC Specification for the Design, Fabrication & Erection of Structural Steel for Buildings, Supplement No. 3 (1974).

Sections shown in **bold face** are "Weight Economy Sections."

S_x in.³	Shape	F'_y ksi	S_x in.³	Shape	F'_y ksi	S_x in.³	Shape	F'_y ksi
114	**W24x55**	**	**57.6**	**W18x35**	**	**21.3**	**W12x19**	**
112	W14x74	**	56.5	W16x36	**	20.9	W8x24	**
112	W10x100	**	54.6	W14x38	**			
111	W21x57	**	54.6	W10x49	53.0	**18.8**	**W10x19**	**
108	W18x60	**	52.0	W8x58	**	18.2	W8x21	**
107	W12x79	**	51.9	W12x40	**			
103	W14x68	**	49.1	W10x45	**	**17.1**	**W12x16**	**
98.5	W10x88	**				16.7	W6x25	**
			48.6	**W14x34**	**	16.2	W10x17	**
98.3	**W18x55**	**				15.2	W8x18	**
97.4	W12x72	52.3	**47.2**	**W16x31**	**			
			45.6	W12x35	**	**14.9**	**W12x14**	54.3
94.5	**W21x50**	**	43.3	W8x48	**	13.8	W10x15	**
92.2	W16x57	**	42.1	W10x39	**	13.4	W6x20	**
92.2	W14x61	**				11.8	W8x15	**
			42.0	**W14x30**	55.3			
88.9	**W18x50**	**				**10.9**	**W10x12**	47.5
87.9	W12x65	43.0	**38.6**	**W12x30**	**	10.2	W6x16	**
85.9	W10x77	**				10.2	W5x19	**
			38.4	**W16x26**	**	9.91	W8x13	**
81.6	**W21x44**	**	35.5	W8x40	**	9.72	W6x15	31.8
81.0	W16x50	**				9.63	M5x18.9	**
78.8	W18x46	**	**35.3**	**W14x26**	**	8.51	W5x16	**
78.0	W12x58	**	35.0	W10x33	50.5			
77.8	W14x53	**				**7.81**	**W8x10**	45.8
75.7	W10x68	**	**33.4**	**W12x26**	57.9	7.31	W6x12	**
72.7	W16x45	**	32.4	W10x30	**			
70.6	W12x53	55.9	31.2	W8x35	**	**5.56**	**W6x9**	50.3
70.3	W14x48	**				5.46	W4x13	**
			29.0	**W14x22**	**			
68.4	**W18x40**	**	27.9	W10x26	**			
66.7	W10x60	**	27.5	W8x31	50.0			
64.7	**W16x40**	**	**25.4**	**W12x22**	**			
64.7	W12x50	**	24.3	W8x28	**			
62.7	W14x43	**						
60.4	W8x67	**	**23.2**	**W10x22**	**			
60.0	W10x54	**						
58.1	W12x45	**						

**Theoretical maximum yield stress exceeds 60 ksi.

MOMENT OF INERTIA
selection table

 $_X$

Sections shown in **bold face** are "Weight Economy Sections."

Moment of Inertia, I_x in.⁴	Shape	Moment of Inertia, I_x in.⁴	Shape	Moment of Inertia, I_x in.⁴	Shape	Moment of Inertia, I_x in.⁴	Shape	Moment of Inertia, I_x in.⁴	Shape	Moment of Inertia, I_x in.⁴	Shape
20300	**W36x300**	**5900**	**W33x118**	**2100**	**W24x76**	843	W21x44	301	W16x26	53.8	W10x12
18900	**W36x280**	5770	W30x132	2070	W21x93	833	W12x96	291	W14x30	53.4	W6x25
17300	**W36x260**	5630	W27x146	1910	W18x106	800	W18x50	285	W12x35	48.0	W8x15
16100	**W36x245**	5440	W14x370	1900	W14x159	796	W14x74	272	W10x49	41.4	W6x20
15000	**W36x230**	5360	W30x124	1890	W12x190	758	W16x57	272	W8x67	39.6	W8x13
14300	W14x730	5170	W24x162	**1830**	**W24x68**	740	W12x87	248	W10x45	32.1	W6x16
14200	W33x241	**4930**	**W30x116**	1830	W21x83	723	W14x68				
13200	**W36x210**	4900	W14x342	1750	W18x97	716	W10x112	**245**	**W14x26**	**30.8**	**W8x10**
12800	W33x221	4580	W24x146	1710	W14x145	712	W18x46	238	W12x30	29.1	W6x15
12400	W14x665	**4470**	**W30x108**	1650	W12x170	662	W12x79	228	W8x58	26.2	W5x19
12100	**W36x194**	4330	W14x311	1600	W21x73	659	W16x50	209	W10x39	24.1	M5x18.9
11500	W33x201	4090	W27x114	**1550**	**W24x62**	640	W14x61			22.1	W6x12
11300	**W36x182**	4020	W24x131	1530	W18x86	623	W10x100	**204**	**W12x26**	21.3	W5x16
10800	W14x605	**3990**	**W30x99**	1530	W14x132	**612**	**W18x40**				
10500	**W36x170**	3840	W14x283	1490	W16x100	597	W12x72	**199**	**W14x22**	**16.4**	**W6x9**
10300	W30x211	3630	W21x147	1480	W21x68	586	W16x45	184	W8x48	11.3	W4x13
9750	**W36x160**	3620	W27x102	1430	W12x152	541	W14x53	170	W10x30		
9430	W14x550	3540	W24x117	1380	W14x120	534	W10x88	170	W10x33		
9170	W30x191	3400	W14x257	**1350**	**W24x55**	533	W12x65				
9040	**W36x150**	**3270**	**W27x94**	1330	W21x62	**518**	**W16x40**	**156**	**W12x22**		
8210	W14x500	3220	W21x132	1330	W18x76	485	W14x48	146	W8x40		
8200	W30x173	3100	W24x104	1300	W16x89	**510**	**W18x35**	144	W10x26		
8160	W33x152	3010	W14x233	1240	W14x109	485	W14x48				
7800	**W36x135**	2960	W21x122	1240	W12x136	475	W12x58	**130**	**W12x19**		
7450	W33x141	**2850**	**W27x84**	1170	W21x57	455	W10x77	127	W8x35		
7190	W14x455	2700	W24x94	1170	W18x71	448	W16x36	118	W10x22		
6990	W27x178	2670	W21x111	1110	W16x77	428	W14x43	110	W8x31		
6710	**W33x130**	2660	W14x211	1110	W14x99	425	W12x53	**103**	**W12x16**		
6600	W14x426	2420	W21x101	1070	W18x65	394	W12x50	98.0	W8x28		
6280	W27x161	2400	W14x193	1070	W12x120	394	W10x68	96.3	W10x19		
6000	W14x398	**2370**	**W24x84**	999	W14x90	385	W14x38				
		2190	W18x119	**984**	**W21x50**	**375**	**W16x31**	**88.6**	**W12x14**		
		2140	W14x176	984	W18x60	350	W12x45	82.8	W8x24		
				954	W16x67	341	W10x60	81.9	W10x17		
				933	W12x106	340	W14x34	75.3	W8x21		
				890	W18x55	310	W12x40	68.9	W10x15		
				882	W14x82	303	W10x54	61.9	W8x18		

PLASTIC SECTION MODULUS

These tables are in accordance with the AISC Specification for the Design, Fabrication & Erection of Structural Steel for Buildings, Supplement No. 3 (1974).

Sections shown in **bold face** are "Weight Economy Sections."

Plastic Section Modulus Z_x	Shape	Cross-Sectional Area A	$b_f/2t_f$	d/t_w	Axis X-X r_x	Axis Y-Y r_y	Yield Stress F_y (KSI) 36 M_p	36 P_y	50 M_p	50 P_y
in.³		in.²			in.	in.	ft-k	kips	ft-k	kips
1260	**W36x300**	88.3	4.96	38.9	15.2	3.83	3780	3180	5250	4420*
1170	**W36x280**	82.4	5.29	41.3	15.1	3.81	3510	2970	4880	4120*
1080	**W36x260**	76.5	5.75	43.2	15.0	3.78	3240	2750*	4500	3830*
1010	**W36x245**	72.1	6.11	45.1	15.0	3.75	3030	2600*	4210	3610*
943	**W36x230**	67.6	6.54	47.2	14.9	3.73	2830	2430*	3930	3380*
939	W33x241	70.9	5.66	41.2	14.1	3.63	2820	2550	3910	3540*
869	W14x426	125	2.75	9.96	7.26	4.34	2610	4500	3620	6250
855	**W33x221**	65.0	6.20	43.8	14.1	3.59	2570	2340*	3560	3250*
833	**W36x210**	61.8	4.48	44.2	14.6	2.58	2500	2220*	3470	3090*
801	W14x398	117	2.92	10.3	7.16	4.31	2400	4210	3340	5850
772	**W33x201**	59.1	6.85	47.1	14.0	3.56	2320	2130*	3220	2960*
767	**W36x194**	57.0	4.81	47.7	14.6	2.56	2300	2050*	3200	2850*
749	W30x211	62.0	5.74	39.9	12.9	3.49	2250	2230	3120	3100*
736	W14x370	109	3.10	10.8	7.07	4.27	2210	3920	3070	5450
718	**W36x182**	53.6	5.12	50.1	14.5	2.55	2150	1930*	2990	2680*
673	W30x191	56.1	6.35	43.2	12.8	3.46	2020	2020*	2800	2810*
672	W14x342	101	3.31	11.4	6.98	4.24	2020	3640	2800	5050
668	**W36x170**	50.0	5.47	53.2	14.5	2.53	2000	1800*	2780	2500*
624	**W36x160**	47.0	5.88	55.4	14.4	2.50	1870	1690*	2600	2350*
605	W30x173	50.8	7.04	46.5	12.7	3.43	1820	1830*	**	**
603	W14x311	91.4	3.59	12.1	6.88	4.20	1810	3290	2510	4570
581	**W36x150**	44.2	6.37	57.4	14.3	2.47	1740	1590*	2420	2210*
567	W27x178	52.3	5.92	38.4	11.6	3.26	1700	1880	2360	2620*
559	W33x152	44.7	5.48	52.7	13.5	2.47	1680	1610*	2330	2230*
542	W14x283	83.3	3.89	13.0	6.79	4.17	1630	3000	2260	4170

*The section should be checked for d/t_w to determine whether it can be used in plastic design (see provisions of AISC Specification Sect. 2.7).

**The shape exceeds width-thickness limitations for plastic design at the indicated yield stress.

PLASTIC SECTION MODULUS

Sections shown in **bold face** are "Weight Economy Sections."

Plastic Section Modulus Z_x	Shape	Cross-Sectional Area A	$b_f/2t_f$	d/t_w	Axis X-X r_x	Axis Y-Y r_y	Yield Stress F_y (KSI)			
							36		50	
							M_p	P_y	M_p	P_y
in.³		in.²			in.	in.	ft-k	kips	ft-k	kips
514	**W33x141**	41.6	6.01	55.0	13.4	2.43	1540	1500*	2140	2080*
512	W27x161	47.4	6.49	41.8	11.5	3.24	1540	1710	2130	2370*
509	**W36x135**	39.7	7.56	59.3	14.0	2.38	1530	1430*	**	**
487	W14x257	75.6	4.23	13.9	6.71	4.13	1460	2720	2030	3780
468	W24x162	47.7	5.31	35.5	10.4	3.05	1400	1720	1950	2380
467	**W33x130**	38.3	6.73	57.1	13.2	2.39	1400	1380*	1950	1920*
461	W27x146	42.9	7.16	45.3	11.4	3.21	1380	1540*	**	**
437	W30x132	38.9	5.27	49.3	12.2	2.25	1310	1400*	1820	1940*
436	W14x233	68.5	4.62	15.0	6.63	4.10	1310	2470	1820	3430
418	W24x146	43.0	5.92	38.1	10.3	3.01	1250	1550	1740	2150*
415	**W33x118**	34.7	7.76	59.7	13.0	2.32	1250	1250*	**	**
408	W30x124	36.5	5.65	51.6	12.1	2.23	1220	1310*	1700	1830*
390	W14x211	62.0	5.06	16.0	6.55	4.07	1170	2230	1630	3100
378	**W30x116**	34.2	6.17	53.1	12.0	2.19	1130	1230*	1580	1710*
373	W21x147	43.2	5.44	30.6	9.17	2.95	1120	1560	1550	2160
370	W24x131	38.5	6.70	40.5	10.2	2.97	1110	1390	1540	1930*
355	W14x193	56.8	5.45	17.4	6.50	4.05	1070	2040	1480	2840
346	**W30x108**	31.7	6.89	54.7	11.9	2.15	1040	1140*	1440	1580*
343	W27x114	33.5	5.41	47.9	11.0	2.18	1030	1210*	1430	1680*
333	W21x132	38.8	6.01	33.6	9.12	2.93	999	1400	1390	1940
327	W24x117	34.4	7.53	44.1	10.1	2.94	981	1240*	**	**
320	W14x176	51.8	5.97	18.3	6.43	4.02	960	1860	1330	2590
312	**W30x99**	29.1	7.80	57.0	11.7	2.10	936	1050*	**	**
311	W12x190	55.8	3.65	13.6	5.82	3.25	933	2010	1300	2790
307	W21x122	35.9	6.45	36.1	9.09	2.92	921	1290	1280	1790
305	W27x102	30.0	6.03	52.6	11.0	2.15	915	1080*	1270	1500*
289	W24x104	30.6	8.50	48.1	10.1	2.91	867	1100*	**	**
287	W14x159	46.7	6.54	20.1	6.38	4.00	861	1680	1200	2330
279	W21x111	32.7	7.05	39.1	9.05	2.90	837	1180	**	**

*The section should be checked for d/t_w to determine whether it can be used in plastic design (see provisions of AISC Specification Sect. 2.7).

**The shape exceeds width-thickness limitations for plastic design at the indicated yield stress.

PLASTIC SECTION MODULUS

These tables are in accordance with the AISC Specification for the Design, Fabrication & Erection of Structural Steel for Buildings, Supplement No. 3 (1974).

Sections shown in **bold face** are "Weight Economy Sections."

Plastic Section Modulus Z_x	Shape	Cross-Sectional Area A	$b_f/2t_f$	d/t_w	Axis X-X r_x	Axis Y-Y r_y	Yield Stress F_y (KSI)			
							36		50	
							M_p	P_y	M_p	P_y
in.³		in.²			in.	in.	ft-k	kips	ft-k	kips
278	**W27x94**	27.7	6.70	54.9	10.9	2.12	834	997*	1160	1380*
275	W12x170	50.0	4.03	14.6	5.74	3.22	825	1800	1150	2500
261	W18x119	35.1	5.31	29.0	7.90	2.69	783	1260	1090	1760
260	W14x145	42.7	7.11	21.7	6.33	3.98	780	1540	**	**
254	**W24x94**	27.7	5.18	47.2	9.87	1.98	762	997*	1060	1380*
253	W21x101	29.8	7.68	42.7	9.02	2.89	759	1070	**	**
244	**W27x84**	24.8	7.78	58.1	10.7	2.07	732	893*	**	**
243	W12x152	44.7	4.46	15.8	5.66	3.19	729	1610	1010	2230
234	W14x132	38.8	7.15	22.7	6.28	3.76	702	1400	**	**
230	W18x106	31.1	5.96	31.7	7.84	2.66	690	1120	958	1560
224	**W24x84**	24.7	5.86	51.3	9.79	1.95	672	889*	933	1230*
221	W21x93	27.3	4.53	37.3	8.70	1.84	663	983	921	1370*
214	W12x136	39.9	4.96	17.0	5.58	3.16	642	1440	892	1990
212	W14x120	35.3	7.80	24.5	6.24	3.74	636	1270	**	**
211	W18x97	28.5	6.41	34.7	7.82	2.65	633	1030	879	1430
200	**W24x76**	22.4	6.61	54.4	9.69	1.92	600	806*	833	1120*
198	W16x100	29.4	5.29	29.0	7.10	2.52	594	1060	825	1470
196	W21x83	24.3	5.00	41.6	8.67	1.83	588	875	817	1220*
192	W14x109	32.0	8.49	27.3	6.22	3.73	576	1150	**	**
186	W18x86	25.3	7.20	38.3	7.77	2.63	558	911	**	**
186	W12x120	35.3	5.57	18.5	5.51	3.13	558	1270	775	1770
177	**W24x68**	20.1	7.66	57.2	9.55	1.87	531	724*	**	**
175	W16x89	26.2	5.92	31.9	7.05	2.49	525	943	729	1310
172	W21x73	21.5	5.60	46.7	8.64	1.81	516	774*	717	1080*
164	W12x106	31.2	6.17	21.1	5.47	3.11	492	1120	683	1560
163	W18x76	22.3	8.11	42.8	7.73	2.61	489	803	**	**
160	**W21x68**	20.0	6.04	49.1	8.60	1.80	480	720*	667	1000*

*The section should be checked for d/t_w to determine whether it can be used in plastic design (see provisions of AISC Specification Sect. 2.7).

*The shape exceeds width-thickness limitations for plastic design at the indicated yield stress.

PLASTIC SECTION MODULUS

Sections shown in **bold face** are "Weight Economy Sections."

Plastic Section Modulus Z_x	Shape	Cross-Sectional Area A	$b_f/2t_f$	d/t_w	Axis X-X r_x	Axis Y-Y r_y	Yield Stress F_y (KSI)			
							36		50	
							M_p	P_y	M_p	P_y
in.³		in.²			in.	in.	ft-k	kips	ft-k	kips
153	**W24x62**	18.2	5.97	55.2	9.23	1.38	459	655*	638	910*
150	W16x77	22.6	6.77	36.3	7.00	2.47	450	814	625	1130
147	W12x96	28.2	6.76	23.1	5.44	3.09	441	1020	613	1410
147	W10x112	32.9	4.17	15.0	4.66	2.68	441	1180	613	1640
145	W18x71	20.8	4.71	37.3	7.50	1.70	435	749	604	1040*
144	**W21x62**	18.3	6.70	52.5	8.54	1.77	432	659*	600	915*
139	W14x82	24.1	5.92	28.1	6.05	2.48	417	868	579	1210
134	**W24x55**	16.2	6.94	59.7	9.11	1.34	402	583*	558	810*
133	W18x65	19.1	5.06	40.8	7.49	1.69	399	688	554	955*
132	W12x87	25.6	7.48	24.3	5.38	3.07	396	922	**	**
130	W16x67	19.7	7.70	41.3	6.96	2.46	390	709	**	**
130	W10x100	29.4	4.62	16.3	4.60	2.65	390	1060	542	1470
129	W21x57	16.7	5.04	52.0	8.36	1.35	387	601*	538	835*
126	W14x74	21.8	6.41	31.5	6.04	2.48	378	785	525	1090
123	W18x60	17.6	5.44	44.0	7.47	1.69	369	634*	513	880*
119	W12x79	23.2	8.22	26.3	5.34	3.05	357	835	**	**
115	W14x68	20.0	6.97	33.8	6.01	2.46	345	720	479	1000
113	W10x88	25.9	5.18	17.9	4.54	2.63	339	932	471	1290
112	**W18x55**	16.2	5.98	46.4	7.41	1.67	336	583*	467	810*
110	**W21x50**	14.7	6.10	54.8	8.18	1.30	330	529*	458	735*
105	W16x57	16.8	4.98	38.2	6.72	1.60	315	605	438	840*
102	W14x61	17.9	7.75	37.0	5.98	2.45	306	644	**	**
101	**W18x50**	14.7	6.57	50.7	7.38	1.65	303	529*	421	735*
97.6	W10x77	22.6	5.86	20.0	4.49	2.60	293	814	407	1130
95.4	**W21x44**	13.0	7.22	59.0	8.06	1.26	286	468*	**	**
92.0	W16x50	14.7	5.61	42.8	6.68	1.59	276	529	383	735*
90.7	W18x46	13.5	5.01	50.2	7.25	1.29	272	486*	378	675*
87.1	W14x53	15.6	6.11	37.6	5.89	1.92	261	562	363	780*
86.4	W12x58	17.0	7.82	33.9	5.28	2.51	259	612	**	**
85.3	W10x68	20.0	6.58	22.1	4.44	2.59	256	720	355	1000
82.3	W16x45	13.3	6.23	46.8	6.65	1.57	247	479*	343	665*

*The section should be checked for d/t_w to determine whether it can be used in plastic design (see provisions of AISC Speciifcation Sect. 2.7).

**The shape exceeds width-thickness limitations for plastic design at the indicated yield stress.

PLASTIC SECTION MODULUS

These tables are in accordance with the AISC Specification for the Design, Fabrication & Erection of Structural Steel for Buildings, Supplement No. 3 (1974).

Sections shown in **bold face** are "Weight Economy Sections."

Plastic Section Modulus Z_x	Shape	Cross-Sectional Area A	$b_f/2t_f$	d/t_w	Axis X-X r_x	Axis Y-Y r_y	Yield Stress F_y (KSI) 36		50	
							M_p	P_y	M_p	P_y
in.³		in.²			in.	in.	ft-k	kips	ft-k	kips
78.4	**W18x40**	11.8	5.73	56.8	7.21	1.27	235	425*	327	590*
78.4	W14x48	14.1	6.75	40.6	5.85	1.91	235	508	327	705*
74.6	W10x60	17.6	7.41	24.3	4.39	2.57	224	634	**	**
72.9	**W16x40**	11.8	6.93	52.5	6.63	1.57	219	425*	304	590*
72.4	W12x50	14.7	6.31	32.9	5.18	1.96	217	529	302	735
70.2	W8x67	19.7	4.43	15.8	3.72	2.12	211	709	292	985
69.6	W14x43	12.6	7.54	44.8	5.82	1.89	209	454*	**	**
66.6	W10x54	15.8	8.15	27.3	4.37	2.56	200	569	**	**
66.5	**W18x35**	10.3	7.06	59.0	7.04	1.22	200	371*	**	**
64.7	W12x45	13.2	7.00	36.0	5.15	1.94	194	475	270	660
64.0	W16x36	10.6	8.12	53.8	6.51	1.52	192	382*	**	**
61.5	W14x38	11.2	6.57	45.5	5.88	1.55	185	403*	256	560*
59.8	W8x58	17.1	5.07	17.2	3.65	2.10	179	616	249	855
57.5	W12x40	11.8	7.77	40.5	5.13	1.93	173	425	**	**
54.9	W10x45	13.3	6.47	28.9	4.33	2.01	165	479	229	665
54.6	**W14x34**	10.0	7.41	49.1	5.83	1.53	164	360*	**	**
54.0	**W16x31**	9.12	6.28	57.7	6.41	1.17	162	328*	225	456*
51.2	W12x35	10.3	6.31	41.7	5.25	1.54	154	371	213	515*
49.0	W8x48	14.1	5.92	21.3	3.61	2.08	147	508	204	705
46.8	W10x39	11.5	7.53	31.5	4.27	1.98	140	414	**	**
44.2	**W16x26**	7.68	7.97	62.8	6.26	1.12	133	276*	**	**
43.1	W12x30	8.79	7.41	47.5	5.21	1.52	129	316*	**	**
40.2	**W14x26**	7.69	5.98	54.5	5.65	1.08	121	277*	168	384*
39.8	W8x40	11.7	7.21	22.9	3.53	2.04	119	421	**	**
36.6	W10x30	8.84	5.70	34.9	4.38	1.37	110	318	153	442
34.7	W8x35	10.3	8.10	26.2	3.51	2.03	104	371	**	**

*The section should be checked for d/t_w to determine whether it can be used in plastic design (see provisions of AISC Specification Sect. 2.7).

**The shape exceeds width-thickness limitations for plastic design at the indicated yield stress.

PLASTIC SECTION MODULUS

Sections shown in **bold face** are "Weight Economy Sections."

Plastic Section Modulus Z_x	Shape	Cross-Sectional Area A	$b_f/2t_f$	d/t_w	Axis X-X r_x	Axis Y-Y r_y	Yield Stress F (KSI)			
							36		50	
							M_p	P_y	M_p	P_y
in.³		in.²			in.	in.	ft-k	kips	ft-k	kips
33.2	**W14x22**	6.49	7.46	59.7	5.54	1.04	99.6	234*	**	**
31.3	W10x26	7.61	6.56	39.7	4.35	1.36	93.9	274	130	380*
29.3	**W12x22**	6.48	4.74	47.3	4.91	0.848	87.9	233*	122	324*
27.2	W8x28	8.25	7.03	28.3	3.45	1.62	81.6	297	**	**
26.0	**W10x22**	6.49	7.99	42.4	4.27	1.33	78.0	234	**	**
24.7	**W12x19**	5.57	5.72	51.7	4.82	0.822	74.1	201*	103	278*
23.2	W8x24	7.08	8.12	32.4	3.42	1.61	69.6	255	**	**
21.6	**W10x19**	5.62	5.09	41.0	4.14	0.874	64.8	202	90.0	281*
20.4	W8x21	6.16	6.59	33.1	3.49	1.26	61.2	222	85.0	308
20.1	**W12x16**	4.71	7.53	54.5	4.67	0.773	60.3	170*	**	**
18.9	W6x25	7.34	6.68	19.9	2.70	1.52	56.7	264	78.8	367
18.7	W10x17	4.99	6.08	42.1	4.05	0.845	56.1	180	77.9	250*
17.0	W8x18	5.26	7.95	35.4	3.43	1.23	51.0	189	**	**
16.0	**W10x15**	4.41	7.41	43.4	3.95	0.810	48.0	159*	**	**
14.9	W6x20	5.87	8.25	23.8	2.66	1.50	44.7	211	**	**
13.6	**W8x15**	4.44	6.37	33.1	3.29	0.876	40.8	160	56.7	222
11.7	W6x16	4.74	4.98	24.2	2.60	0.967	35.1	171	48.8	237
11.6	W5x19	5.54	5.85	19.1	2.17	1.28	34.8	199	48.3	277
11.4	**W8x13**	3.84	7.84	34.7	3.21	0.843	34.2	138	**	**
11.0	M5x18.9	5.55	6.01	15.8	2.08	1.19	33.0	200	45.8	278
9.59	W5x16	4.68	6.94	20.9	2.13	1.27	28.8	168	40.0	234
8.30	**W6x12**	3.55	7.14	26.2	2.49	0.918	24.9	128	**	**
6.28	W4x13	3.83	5.88	14.9	1.72	1.00	18.8	138	26.2	192

*The section should be checked for d/t_w to determine whether it can be used in plastic design (see provisions of AISC Specification Sect. 2.7).

**The shape exceeds width-thickness limitations for plastic design at the indicated yield stress.

Lc and Lu tables
UNBRACED LENGTHS

These tables are in accordance with the AISC Specification for the
Design, Fabrication & Erection of Structural Steel for Buildings,
Supplement No. 3 (1974).

Shape	Fy=36 ksi		Fy=50 ksi		Shape	Fy=36 ksi		Fy=50 ksi	
	Lc	Lu	Lc	Lu		Lc	Lu	Lc	Lu
	ft	ft	ft	ft		ft	ft	ft	ft
W36 x 300	17.6	35.3	14.9	25.4	W24 x 162	13.7	29.3	11.6	21.1
280	17.5	33.0	14.9	23.8	146	13.6	26.3	11.6	18.9
260	17.5	30.4	14.8	21.9	131	13.6	23.3	11.5	16.8
245	17.4	28.6	14.8	20.6	117	13.5	20.8	11.5	14.9
230	17.4	26.8	14.8	19.3	104	13.5	18.4	11.4	13.2
W36 x 210	12.9	20.9	10.9	15.0	W24 x 94	9.6	15.1	8.1	10.9
194	12.8	19.4	10.9	13.9	84	9.5	13.3	8.1	9.6
182	12.7	18.2	10.8	13.1	76	9.5	11.8	8.1	8.6
170	12.7	16.9	10.8	12.2	68	9.5	10.2	7.4	8.5
160	12.7	15.7	10.7	11.4					
150	12.6	14.5	10.5	11.3	W24 x 62	7.4	8.1	5.8	6.4
135	12.3	13.0	8.9	11.0	55	6.9	7.5	5.0	6.3
W33 x 241	16.7	30.1	14.2	21.7	W21 x 147	13.2	30.2	11.2	21.7
221	16.7	27.5	14.2	19.8	132	13.1	27.3	11.1	19.7
201	16.6	24.9	14.1	17.9	122	13.1	25.4	11.1	18.3
					111	13.0	23.2	11.1	16.7
W33 x 152	12.2	16.9	10.4	12.1	101	13.0	21.3	11.0	15.3
141	12.2	15.4	10.3	11.1					
130	12.1	13.8	9.9	10.9	W21 x 93	8.9	16.8	7.5	12.1
118	12.0	12.6	8.6	10.7	83	8.8	15.1	7.5	10.9
					73	8.8	13.4	7.4	9.6
W30 x 211	15.9	29.7	13.5	21.4	68	8.7	12.4	7.4	8.9
191	15.9	26.9	13.5	19.4	62	8.7	11.2	7.4	8.0
173	15.8	24.3	13.4	17.5					
					W21 x 57	6.9	9.4	5.9	6.7
W30 x 132	11.1	16.1	9.4	11.6	50	6.9	7.8	5.6	6.0
124	11.1	15.0	9.4	10.8	44	6.6	7.0	4.7	5.9
116	11.1	13.8	9.4	9.9					
108	11.1	12.4	8.9	9.8	W18 x 119	11.9	29.1	10.1	21.0
99	10.9	11.4	7.9	9.7	106	11.8	26.0	10.0	18.7
					97	11.8	24.1	10.0	17.4
W27 x 178	14.9	27.9	12.6	20.1	86	11.7	21.5	9.9	15.5
161	14.8	25.4	12.6	18.3	76	11.6	19.1	9.9	13.7
146	14.7	23.0	12.5	16.6					
					W18 x 71	8.1	15.5	6.8	11.2
W27 x 114	10.6	15.9	9.0	11.4	65	8.0	14.4	6.8	10.3
102	10.6	14.2	9.0	10.2	60	8.0	13.3	6.8	9.6
94	10.5	12.8	8.9	9.5	55	7.9	12.1	6.7	8.7
84	10.5	11.1	8.0	9.4	50	7.9	11.0	6.7	7.9

L_c and L_u tables
UNBRACED LENGTHS

Shape	F_y=36 ksi		F_y=50 ksi		Shape	F_y=36 ksi		F_y=50 ksi	
	L_c	L_u	L_c	L_u		L_c	L_u	L_c	L_u
	ft	ft	ft	ft		ft	ft	ft	ft
W18 x 46	6.4	9.4	5.4	6.8	W14 x 132	15.5	47.9	13.2	34.5
40	6.3	8.2	5.4	5.9	120	15.5	44.1	13.1	31.7
35	6.3	6.7	4.8	5.6	109	15.4	40.6	13.1	29.2
					99	15.4	37.1	13.0	26.7
W16 x 100	11.0	28.0	9.3	20.2	90	15.3	34.0	13.0	24.5
89	10.9	25.1	9.3	18.0					
77	10.9	21.9	9.2	15.8	W14 x 82	10.7	28.0	9.1	20.2
67	10.8	19.3	9.2	13.9	74	10.6	25.8	9.0	18.6
					68	10.6	23.8	9.0	17.2
W16 x 57	7.5	14.3	6.4	10.	61	10.6	21.5	9.0	15.5
50	7.5	12.7	6.3	9.1					
45	7.4	11.4	6.3	8.2	W14 x 53	8.5	17.7	7.2	12.7
40	7.4	10.2	6.3	7.4	48	8.5	16.0	7.2	11.5
36	7.4	8.8	6.3	6.7	43	8.4	14.4	7.2	10.3
W16 x 31	5.8	7.1	4.9	5.2	W14 x 38	7.1	11.4	6.1	8.2
26	5.6	6.0	4.0	5.1	34	7.1	10.2	6.0	7.3
					30	7.1	8.7	6.0	6.5
W14 x 730	18.9	181.4	16.0	130.6					
665	18.6	170.7	15.8	122.9	W14 x 26	5.3	7.0	4.5	5.1
605	18.4	160.3	15.6	115.4	22	5.3	5.6	4.1	4.7
550	18.2	150.3	15.4	108.2					
500	18.0	140.6	15.2	101.2	W12 x 190	13.4	70.8	11.3	51.0
455	17.8	131.5	15.1	94.7	170	13.3	64.7	11.3	46.6
					152	13.2	59.0	11.2	42.5
W14 x 426	17.6	125.6	15.0	90.5	13f	13.1	53.5	11.1	38.5
398	17.5	119.5	14.9	86.0	120	13.0	48.0	11.0	34.6
370	17.4	113.2	14.8	81.5	106	12.9	43.5	10.9	31.3
342	17.3	106.7	14.7	76.8	96	12.8	39.9	10.9	28.7
311	17.1	99.2	14.5	71.4	87	12.8	36.3	10.9	26.1
283	17.0	92.2	14.4	66.4	79	12.8	33.2	10.8	23.9
257	16.9	85.4	14.3	61.5	72	12.7	30.5	10.8	22.0
233	16.8	78.9	14.2	56.8	65	12.7	27.7	10.7	20.0
211	16.7	72.6	14.2	52.3					
193	16.6	67.7	14.1	48.7	W12 x 58	10.6	24.3	9.0	17.5
176	16.5	62.4	14.0	44.9	53	10.6	22.1	9.0	15.9
159	16.4	57.2	13.9	41.2					
145	16.4	52.9	13.9	38.1	W12 x 50	8.5	19.6	7.2	14.1
					45	8.5	17.8	7.2	12.8
					40	8.4	16.0	7.2	11.5

L_c and L_u tables
UNBRACED LENGTHS

These tables are in accordance with the AISC Specification for the
Design, Fabrication & Erection of Structural Steel for Buildings,
Supplement No. 3 (1974).

Shape	F_y=36 ksi		F_y=50 ksi		Shape	F_y=36 ksi		F_y=50 ksi	
	L_c	L_u	L_c	L_u		L_c	L_u	L_c	L_u
	ft	ft	ft	ft		ft	ft	ft	ft
W12 x 35	6.9	12.6	5.9	9.1	W8 x 21	5.6	11.8	4.7	8.5
30	6.9	10.8	5.8	7.7	18	5.5	9.9	4.7	7.1
26	6.9	9.3	5.8	6.7	W8 x 15	4.2	7.2	3.6	5.2
W12 x 22	4.3	6.4	3.6	4.6	13	4.2	5.9	3.6	4.3
19	4.2	5.3	3.6	3.8	10	4.2	4.7	3.4	3.7
16	4.1	4.3	2.9	3.6	W6 x 25	6.4	20.1	5.4	14.5
14	3.5	4.2	2.5	3.6	20	6.4	16.4	5.4	11.8
W10 x 112	11.0	53.1	9.3	38.2	15	6.3	12.0	5.4	8.7
100	10.9	48.3	9.3	34.8	W6 x 16	4.3	12.0	3.6	8.7
88	10.8	43.4	9.2	31.2	12	4.2	8.6	3.6	6.2
77	10.8	38.7	9.1	27.9	9	4.2	6.6	3.5	4.8
68	10.7	34.7	9.1	25.0	W5 x 19	5.3	19.4	4.5	14.0
60	10.6	31.1	9.0	22.4	16	5.3	16.6	4.5	12.0
54	10.6	28.3	9.0	20.4	W4 x 13	4.3	15.6	3.6	11.2
49	10.6	26.0	9.0	18.7	M5 x 18.9	5.3	19.3	4.5	13.9
W10 x 45	8.5	22.8	7.2	16.4					
39	8.4	19.8	7.2	14.2					
33	8.4	16.5	7.1	11.9					
W10 x 30	6.1	13.1	5.2	9.4					
26	6.1	11.4	5.2	8.2					
22	6.1	9.4	5.2	6.8					
W10 x 19	4.2	7.2	3.6	5.2					
17	4.2	6.1	3.6	4.4					
15	4.2	5.0	3.6	3.7					
12	3.9	4.3	2.8	3.6					
W8 x 67	8.7	39.8	7.4	28.7					
58	8.7	35.2	7.4	25.4					
48	8.6	30.3	7.3	21.8					
40	8.5	25.4	7.2	18.3					
35	8.5	22.6	7.2	16.3					
31	8.4	20.1	7.2	14.5					
W8 x 28	6.9	17.5	5.9	12.6					
24	6.9	15.2	5.8	10.9					

DATA SHEET A24

ALLOWABLE STRESS
FOR COMPRESSION MEMBERS OF 36 KSI SPECIFIED YIELD STRESS STEEL

$F_y = 36 \text{ ksi}$

Main and Secondary Members Kl/r not over 120						Main Members Kl/r 121 to 200				Secondary Members[a] l/r 121 to 200			
$\frac{Kl}{r}$	F_a (ksi)	$\frac{Kl}{r}$	F_a (ksi)	$\frac{Kl}{r}$	F_a (ksi)	$\frac{Kl}{r}$	F_a (ksi)	$\frac{Kl}{r}$	F_a (ksi)	$\frac{l}{r}$	F_{as} (ksi)	$\frac{l}{r}$	F_{as} (ksi)
1	21.56	41	19.11	81	15.24	121	10.14	161	5.76	121	10.19	161	7.25
2	21.52	42	19.03	82	15.13	122	9.99	162	5.69	122	10.09	162	7.20
3	21.48	43	18.95	83	15.02	123	9.85	163	5.62	123	10.00	163	7.16
4	21.44	44	18.86	84	14.90	124	9.70	164	5.55	124	9.90	164	7.12
5	21.39	45	18.78	85	14.79	125	9.55	165	5.49	125	9.80	165	7.08
6	21.35	46	18.70	86	14.67	126	9.41	166	5.42	126	9.70	166	7.04
7	21.30	47	18.61	87	14.56	127	9.26	167	5.35	127	9.59	167	7.00
8	21.25	48	18.53	88	14.44	128	9.11	168	5.29	128	9.49	168	6.96
9	21.21	49	18.44	89	14.32	129	8.97	169	5.23	129	9.40	169	6.93
10	21.16	50	18.35	90	14.20	130	8.84	170	5.17	130	9.30	170	6.89
11	21.10	51	18.26	91	14.09	131	8.70	171	5.11	131	9.21	171	6.85
12	21.05	52	18.17	92	13.97	132	8.57	172	5.05	132	9.12	172	6.82
13	21.00	53	18.08	93	13.84	133	8.44	173	4.99	133	9.03	173	6.79
14	20.95	54	17.99	94	13.72	134	8.32	174	4.93	134	8.94	174	6.76
15	20.89	55	17.90	95	13.60	135	8.19	175	4.88	135	8.86	175	6.73
16	20.83	56	17.81	96	13.48	136	8.07	176	4.82	136	8.78	176	6.70
17	20.78	57	17.71	97	13.35	137	7.96	177	4.77	137	8.70	177	6.67
18	20.72	58	17.62	98	13.23	138	7.84	178	4.71	138	8.62	178	6.64
19	20.66	59	17.53	99	13.10	139	7.73	179	4.66	139	8.54	179	6.61
20	20.60	60	17.43	100	12.98	140	7.62	180	4.61	140	8.47	180	6.58
21	20.54	61	17.33	101	12.85	141	7.51	181	4.56	141	8.39	181	6.56
22	20.48	62	17.24	102	12.72	142	7.41	182	4.51	142	8.32	182	6.53
23	20.41	63	17.14	103	12.59	143	7.30	183	4.46	143	8.25	183	6.51
24	20.35	64	17.04	104	12.47	144	7.20	184	4.41	144	8.18	184	6.49
25	20.28	65	16.94	105	12.33	145	7.10	185	4.36	145	8.12	185	6.46
26	20.22	66	16.84	106	12.20	146	7.01	186	4.32	146	8.05	186	6.44
27	20.15	67	16.74	107	12.07	147	6.91	187	4.27	147	7.99	187	6.42
28	20.08	68	16.64	108	11.94	148	6.82	188	4.23	148	7.93	188	6.40
29	20.01	69	16.53	109	11.81	149	6.73	189	4.18	149	7.87	189	6.38
30	19.94	70	16.43	110	11.67	150	6.64	190	4.14	150	7.81	190	6.36
31	19.87	71	16.33	111	11.54	151	6.55	191	4.09	151	7.75	191	6.35
32	19.80	72	16.22	112	11.40	152	6.46	192	4.05	152	7.69	192	6.33
33	19.73	73	16.12	113	11.26	153	6.38	193	4.01	153	7.64	193	6.31
34	19.65	74	16.01	114	11.13	154	6.30	194	3.97	154	7.59	194	6.30
35	19.58	75	15.90	115	10.99	155	6.22	195	3.93	155	7.53	195	6.28
36	19.50	76	15.79	116	10.85	156	6.14	196	3.89	156	7.48	196	6.27
37	19.42	77	15.69	117	10.71	157	6.06	197	3.85	157	7.43	197	6.26
38	19.35	78	15.58	118	10.57	158	5.98	198	3.81	158	7.39	198	6.24
39	19.27	79	15.47	119	10.43	159	5.91	199	3.77	159	7.34	199	6.23
40	19.19	80	15.36	120	10.28	160	5.83	200	3.73	160	7.29	200	6.22

[a] K taken as 1.0 for secondary members.

Note: $C_c = 126.1$

From THE *MANUAL OF STEEL CONSTRUCTION, 8th ED*. Reprinted with permission
of the American Institute of Steel Construction.

209

ANSWERS TO PROBLEMS

Problem 4.1: *(a)* W = 37.7^k;
(b) P = 23.5^k; *(c)* P = 44.8^k

Problem 4.2: *(a)* W21 × 44; *(b)* W21 ×
44; *(c)* W16 × 40; *(d)* W21 × 44;
(e) W21 × 62; *(f)* W21 × 44;
(g) W27 × 94; *(h)* W27 × 84;
(i) W30 × 116; *(j)* W27 × 84;
(k) W24 × 55; *(l)* W27 × 94

Problem 4.3: *(a)* W14 × 34*;
(b) W14 × 34; *(c)* W18 × 35;
(d) W10 × 19*
*supplementary lateral support
recommended

Problem 4.4: *(a)* W12 × 16*;
(b) W12 × 14*; *(c)* W12 × 19;
(d) W12 × 14*; *(e)* W16 × 31;
(f) W12 × 16*
*lateral bracing recommended for
cantilevers

Problem 4.5: **(a):** *B1*, W6 × 9; *B2*,
W14 × 22; *B3*, W14 × 26; *B4*,
W21 × 44; *B5*, W14 × 22 (deflection);
G1, W18 × 40; *G2*, W16 × 31; *G3*,
W21 × 44. **(b):** deflection controls all
parts; *B1*, W12 × 22; *B2*, W18 × 40;
G1, W30 × 108

Problem 4.6: **(1):** *beam (a)*, W14 × 30;
beam (b), W14 × 34. **(2):** midspan
Δ = 0.91"; left cantilever Δ = −0.63";
right cantilever Δ = −0.50".
(3): midspan Δ = 0.15"; left cantilever
Δ = 0.18"; right cantilever Δ = 0.10"

Problem 4.7: *beam (a)*, W16 × 40 with
bracing; *beam (b)*, W21 × 44

Problem 4.8: *beam (a)*, W21 × 44 (brace
center span); *beam (b)*, W24 × 55
(brace center span)

Problem 4.9: **(a):** *B1*, W21 × 44; *B2*,
W21 × 44; *B3*, W10 × 12. **(b):** *B1*,
W = 57.2^k; *B2*, W = 60.8^k; *B3*,
W = 19.3^k. **(c):** Δ = 0.31"; Δ = 0.57"

Problem 4.10: **(1):** M = 162$^{'k}$;
(2): W = 65^k; **(3):** Δ = 0.32";
(4): W21 × 44, Δ = 0.47"

Problem 4.11: **(1):** concrete f = 0.92 ksi;
steel f = 24.7 ksi; **(2):** f = 35.4 ksi

Problem 4.12: **(1):** W36 × 182;
Δ = 2.31"; **(2):** 92"; **(3):** concrete
f = 1.04 ksi; steel f = 23.97 ksi;
Δ = 1.53"

Problem 5.1: P $= 90.2^k$ P $= 60.8^k$
                         P $= 60.3^k$ P $= 196.7^k$
                         P $= 68.3^k$ P $= 24.6^k$
                         P $= 39.6^k$ P $= 145.8^k$
                         P $= 191.5^k$ P $= 170.6^k$

Problem 5.2: (1): W10 $\times$ 45;
   (2): W12 $\times$ 58; (3): W8 $\times$ 35

Problem 5.3: W8 $\times$ 21

Problem 5.4: P $= 385.8^k$

Problem 5.5: P $= 244.2^k$

Problem 5.6: W12 $\times$ 58

Problem 5.7: (1): P $= 133.4^k$;
   (2): P $= 61.9^k$

Problem 5.8: W14 $\times$ 61

Problem 5.9: W10 $\times$ 39

Problem 5.10: W14 $\times$ 48

Problem 7.1: A $= 1.63$ in^2

Problem 7.2: (1): *4th*, F $= 2.17^k$; *3rd*,
   F $= 5.42^k$; *2nd*, F $= 8.67^k$; *1st*,
   F $= 10.83^k$; (2): W8 $\times$ 24; (3): *CI*,
   P $= 6.23^k$

Problem 7.3: A $= 0.46$ in^2

Problem 7.4: (1): *7th*, F $= 2.5^k$; *6th*,
   F $= 7.5^k$; *5th*, F $= 6.25^{k*}$; *4th*,
   F $= 17.5^k$; *3rd*, F $= 22.5^k$k; *2nd*,

F $= 26.25^k$; *1st*, F $= 30^k$; (2): *CI*,
29.25^k

*sloped member from 4th to 5th floor
would be designed as a compression
member for forces from the opposite
direction

Problem 7.5: (a): lower level, interior
column M $= 97.2^{'k}$; (b): 31.3^k;
(c): 28.2^k; (d): W24 $\times$ 68

Problem 8.1: (a): *(a)* W24 $\times$ 62;
   (b) W24 $\times$ 68; *(c)* W21 $\times$ 50;
   (d) W18 $\times$ 50; *(e)* W18 $\times$ 55; (b):
   (a) W18 $\times$ 50; *(b)* W21 $\times$ 68;
   (c) W21 $\times$ 44; *(d)* W18 $\times$ 40;
   (e) W21 $\times$ 50

Problem 8.2: *(a)* W16 $\times$ 40;
   (b) W18 $\times$ 35; *(c)* W18 $\times$ 35;
   (d) W21 $\times$ 44; *(e)* W16 $\times$ 40

Problem 8.3: (1): Z $= 101$ in^3;
   (2): $W_p = 220.6^k$; $P_p = 90.9^k$

Problem 8.4: (1): Z $= 124.5$ in^3;
   (2): $P_p = 112^k$; $W_p = 217.6^k$;
   (3): $P_p = 139.4^k$;

Problem 8.5: (1): $M_p = 225.6^{'k}$;
   $P_p = 61.7^k$; (2): midspan

INDEX

Contents